Great Science Adventures

The World of Insects and Arachnids

by
Dinah Zike
and
Susan Simpson

See where learning takes you.

www.greatscienceadventures.com

Great Science Adventures is a comprehensive project which is projected to include the titles below. Please check our website, www.greatscienceadventures.com, for updates and product availability.

Great Life Science Studies:
- *The World of Plants*
- *The World of Insects and Arachnids*
- *The World of the Human Body*
- *The World of Vertebrates*
- *The World of Biomes*
- *The World of Health*

Great Physical Science Studies:
- *The World of Tools and Technology*
- *The World of Matter and Energy*
- *The World of Light and Sound*
- *The World of Electricity and Magnets*

Great Earth Science Studies:
- *The World of Space*
- *The World of Atmosphere and Weather*
- *The World of Lithosphere / Earth*
- *The World of Hydrosphere / Fresh Water*
- *The World of Hydrosphere / Oceans*
- *The World of Rocks and Minerals*

Copyright © 2001 by:
Common Sense Press
8786 Highway 21
Melrose, FL 32666
(352) 475-5757
www.greatscienceadventures.com

All rights reserved. No part of this book may be reproduced in any form without written permission from Common Sense Press.

Printed in the United States of America
ISBN 1-929683-08-1

The authors and the publisher have made every reasonable effort to ensure that the experiments and activities in this book are safe when performed according to the book's instructions. We assume no responsibility for any damage sustained or caused while performing the activities or experiments in *Great Science Adventures*. We further recommend that students undertake these activities and experiments under the supervision of a teacher, parent, and/or guardian.

Great Science Adventures

Table of Contents

Introduction	.i
How to Use This Program	.iii
1. What are insects?	.10
2. Who studies insects?	.14
3. What are the characteristics of insects?	.18
4. How do insects use their mouths and antennae?	.20
5. How do insects see?	.22
6. What is the form and function of insects?	.26
7. How do insects reproduce?	.30
8. How do insects' respiratory and circulatory systems function?	.34
9. How do insects' digestive and nervous systems function?	.36
10. What is complete metamorphosis?	.40
11. What is incomplete metamorphosis?	.44
12. How do insects defend themselves?	.48
13. What are predatory insects?	.50
14. Where do insects live?	.52
15. Do insects migrate?	.54
16. Why are ants and termites called social insects?	.56
17. Why are honeybees called social insects?	.58
18. What are the differences between butterflies and moths?	.62
19. What are beetles?	.64
20. Are insects helpful or harmful?	.68
21. What are crustaceans?	.72
22. What are arachnids?	.74
23. What are spiders?	.76
24. What are mites and ticks?	.80
Assessment	.82
Lots of Science Library Books	.85
Graphics Pages	.147

Great Science Adventures

Introduction

Great Science Adventures is a unique, highly effective program that is easy to use for teachers as well as students. This book contains 24 lessons. The concepts to be taught are clearly listed at the top of each lesson. Activities, questions, clear directions, and pictures are included to help facilitate learning. Each lesson will take one to three days to complete.

This program utilizes highly effective methods of learning. Students not only gain knowledge of basic science concepts, but also learn how to apply them.

Specially designed *3D Graphic Organizers* are included for use with the lessons. These organizers review the science concepts while adding to your students' understanding and retention of the subject matter.

This *Great Science Adventures* book is divided into four parts:

1) Following this *Introduction* you will find the *How to Use This Program* section. It contains all the information you need to make the program successful. The *How to Use This Program* section also contains instructions for Dinah Zike's *3D Graphic Organizers*. Please take the time to learn the terms and instructions for these learning manipulatives.

2) In the *Teacher's Section,* the numbered lessons include a list of the science concepts to be taught, simple to complex vocabulary words, and activities that reinforce the science concepts. Each activity includes a list of materials needed, directions, pictures, questions, written assignments, and other helpful information for the teacher.

    The *Teacher's Section* also includes enrichment activities, entitled *Experiences, Investigations, and Research.* Alternative assessment suggestions are found at the end of the *Teacher's Section.*

3) The *Lots of Science Library Books* are next. These books are numbered to correlate with the lessons. Each *Lots of Science Library Book* will cover all the concepts included in its corresponding lesson. You may read the *LSLB* to your students, ask them to read the books on their own, or make the books available as research materials. Covers for the books are found at the beginning of the *LSLB* section. (Common Sense Press grants permission for you to photocopy the *Lots of Science Library Books* pages and covers for your students.)

4) *Graphic Pages,* also listed by lesson numbers, provide pictures and graphics that can be used with the activities. They can be duplicated and used on student-made manipulatives, or students may draw their own illustrations. The *Investigative Loop* at the beginning of this section may be photocopied, as well. (Common Sense Press grants permission for you to photocopy the *Graphics Pages* for your students.)

Great Science Adventures

How to Use This Program

This program can be used in a single-level classroom, multilevel classroom, homeschool, co-op group, or science club. Everything you need for a complete insect and arachnid study is included in this book. Intermediate students will need access to basic reference materials.

Take the time to read the entire *How to Use This Program* section and become familiar with the sections of this book that are described in the *Introduction*.

Begin a lesson by reading the *Teacher Pages* for that lesson. Choose the vocabulary words for each student and the activities to complete. Collect the materials you need for these activities.

Introduce each lesson with its corresponding *Lots of Science Library Book* by reading it aloud or asking a student to read it. (The *Lots of Science Library Books* are located after the *Teacher's Section*.)

Discuss the concepts presented in the *Lots of Science Library Book,* focusing on the ones listed in your *Teacher's Section.*

Follow the directions for the activities you have chosen.

How to Use the Multilevel Approach

The lessons in this book include basic content appropriate for grades K–8 at different mastery levels. For example, throughout the teaching process, a first grader will be exposed to a lot of information but would not be expected to retain all of it. In the same lesson, a sixth-grade student will learn all the steps of the process, be able to communicate them in writing, and be able to apply that information to different situations.

In the *Lots of Science Library Books,* the words written in larger type are for all students. The words in smaller type are for upper level students and include more scientific details about the basic content, as well as interesting facts for older learners.

In the activity sections, icons are used to designate the levels of specific writing assignments.

This icon ✎ indicates the Beginning level, which includes the non-reading or early reading student. This level applies mainly to kindergarten and first grade students.

This icon ✎✎ is used for the Primary level. It includes the reading student who is still working to be a fluent reader. This level is designed primarily for second and third graders.

This icon ✎✎✎ denotes the Intermediate level, or fluent reader. This level of activities will usually apply to fourth through eighth grade students.

If you are working with a student in seventh or eighth grade, we recommend using the assignments for the Intermediate level, plus at least one *Experiences, Investigations, and Research* activity per lesson.

No matter what grade level your students are working on, use a level of written work that is appropriate for their reading and writing abilities. It is good for students to review data they already know, learn new data and concepts, and be exposed to advanced information and processes.

Vocabulary Words

Each lesson lists vocabulary words that are used in the content of the lesson. Some of these words will be "too easy" for your students, some will be "too hard," and others will be "just right." The "too easy" words will be used automatically during independent writing assignments. Words that are "too hard" can be used during discussion times. Words that are "just right" can be studied by definition, usage, and spelling. Encourage your students to use these words in their own writing and speaking.

You can encourage beginning students to use their vocabulary words as you reinforce reading instruction and enhance discussions about the topic, and as words to be copied in cooperative, or teacher guided, writing.

Primary and Intermediate students can make a Vocabulary Book for new words. Instructions for making a Vocabulary Book are found on page 3. The Vocabulary Book will contain the word definitions and sentences composed by the student for each word. Students should also be expected to use their vocabulary words in discussions and independent writing assignments. A vocabulary word with an asterisk (*) next to it is designated for Intermediate students only.

Using 3D Graphic Organizers

The *3D Graphic Organizers* provide a format for students of all levels to conceptualize, analyze, review, and apply the concepts of the lesson. The *3D Graphic Organizers* take complicated information and break it down into visual parts so students can better understand the concepts. Most *3D Graphic Organizers* involve writing about the subject matter. Although the content for the levels will generally be the same, assignments and expectations for the levels will vary.

Beginning students may dictate or copy one or two "clue" words about the topic. These students will use the written clues to verbally communicate the science concept. The teacher should provide various ways for the students to restate the concept. This will reinforce the science concept and encourage the students in their reading and higher order thinking skills.

Primary students may write or copy one or two "clue" words and a sentence about the topic. The teacher should encourage students to use vocabulary words when writing these sentences. As students read their sentences and discuss them, they will reinforce the science concept, increasing their fluency in reading and higher order thinking skills.

Intermediate students may write several sentences or a paragraph about the topic. These students are also encouraged to use reference materials to expand their knowledge of the subject. As tasks are completed, students enhance their abilities to locate information, read for content, compose sentences and paragraphs, and increase vocabulary. Encourage these students to use the vocabulary words in a context that indicates understanding of the words' meanings.

Illustrations for the *3D Graphic Organizers* are found on the *Graphics Pages* and are labeled by the lesson number and a letter, such as 5–A. Your students may use these graphics to draw their own pictures, or cut out and glue them directly on their work.

Several of the *3D Graphic Organizers* will be used over a series of lessons. For this reason, you will need a storage system for each student's *3D Graphic Organizers*. A pocket folder or a reclosable plastic bag works well. See page 1 for more information on storing materials.

Investigative Loop™

The *Investigative Loop* is used throughout *Great Science Adventures* to ensure that your labs are effective and practical. Labs give students a context for the application of their science lessons, so that they begin to take ownership of the concepts, increasing understanding as well as retention.

The *Investigative Loop* can be used in any lab. The steps are easy to follow, user friendly, and flexible.

Each *Investigative Loop* begins with a **Question or Concept.** If the lab is designed to answer a question, use a question in this phase. For example, the question could be: "How does an insect see an object?"

If the lab is designed to demonstrate a concept, use a concept statement in this phase, such as: "Insects use scent to locate objects." The lab will demonstrate that fact to the students.

After the **Question or Concept** is formulated, the next phase of the *Investigative Loop* is Research and/or Predictions. Research gives students a foundation for the lab. Having researched the question or concept, students enter the lab with a basis for understanding what they observe. Predictions are best used when the first phase is a question. Predictions can be in the form of a statement, a diagram, or a sequence of events.

 The **Procedure** for the lab follows. This is an explanation of how to set up the lab and any tasks involved in it. A list of materials for the lab may be included in this section or may precede the entire *Investigative Loop*.

Whether the lab is designed to answer a question or demonstrate a concept, the students' **Observations** are of primary importance. Tell the students what to focus upon in their observations. The Observation phase will continue until the lab ends.

 Once observations are made, students must **Record the Data**. Data may be recorded through diagrams or illustrations. Recording quantitative or qualitative observations of the lab is another important activity in this phase. Records may be kept daily for an extended lab or at the beginning and end for a short lab.

Conclusions and/or Applications are completed when the lab ends. Usually the data records will be reviewed before a conclusion can be drawn about the lab. Encourage the students to defend their conclusions by using the data records. Applications are made by using the conclusions to generalize to other situations or by stating how to use the information in daily life.

 Next, **Communicate the Conclusions**. This phase is an opportunity for students to be creative. Conclusions can be communicated through a graph, story, report, video, mock radio show, etc. Students may also participate in a group presentation.

Questions that are asked as the activity proceeds are called **Spark Questions.** Questions that the lab sparks in the minds of the students are important to discuss when the lab ends. The lab itself will answer many of these questions, while others may lead to a new *Investigative Loop*. Assign someone to keep a list of all Spark Questions.

 One lab naturally leads to another. This begins a new *Investigative Loop*. The phase called **New Loop** is a brainstorming time for narrowing the lab down to a new question or concept. When the new lab has been decided upon, the *Investigative Loop* begins again with a new Question or Concept.

Take the time to teach your students to make qualitative and quantitative observations. Qualitative observations involve recording the color, texture, shape, smell, size (such as small, medium, large), or any words that describe the qualities of an object. Quantitative observations involve using a standard unit of measurement to determine the length, width, weight, mass, or volume of an object.

All students will make a Lab Book, in the form of a Pocket Book, to store information about the Investigative Loops. Instructions for making a Pocket Book are found on page 2. Your students will make a new Lab Book as needed to glue side–by–side to the previous one. Instructions can be found in the *Teacher's Section*.

Predictions, data, and conclusions about the *Investigative Loops* are usually written on Lab Record Cards. These can be 3– x 5–inch index cards or paper cut to size.

When you begin an *Investigative Loop*, ask your students to glue or draw the graphic of the experiment on the pocket of the Lab Book. Each *Investigative Loop* is labeled with the lesson number and another number. These numbers are also found on the corresponding graphics. The completed Lab Record Cards will be labeled by Lab Number and placed in the appropriate pocket.

During an *Investigative Loop*, beginning students should be encouraged to discuss their answers to all experiment questions. By discussing the topic, the students will not only learn the science concepts and procedures, but will be able to organize their thinking in a manner that will enhance their writing skills. This discussion time is very important for beginning students and should not be rushed.

After the discussion, work with the students to construct a sentence about the topic. Let them copy the sentence. Students can also write "clue" words to help them remember key points about the experiment and discuss it at a later time.

Primary students should be encouraged to verbalize their answers. By discussing the topic, students will learn the science concepts and procedures and learn to organize their thinking, increasing their ability to use higher–level thinking skills. After the discussion, students can complete the assignment using simple phrases or sentences. Encourage students to share the information they have learned with others, such as parents or friends. This will reinforce the content and skills covered in the lesson.

Even though Intermediate students can write the answers to the lab assignments, the discussion process is very important and should not be skipped. By discussing the experiments, students review the science concepts and procedures as well as organize their thinking for the writing assignments. This allows them to think and write at higher levels. These students should be encouraged to use their vocabulary words in their lab writing assignments.

Design Your Own Experiment

After an *Investigative Loop* is completed, intermediate students have the option to design their own experiments based on that lab. The following procedure should be used for those experiments.

Select a Topic based upon an experience in an *Investigative Loop*, science content, an observation, a high-interest topic, a controversial topic, or a current event.

Discuss the Topic as a class, in student groups, and with knowledgeable professionals.

Read and Research the Topic using the library, the Internet, and hands-on investigations and observations, when possible.

Select a Question that can be investigated and answered using easily obtained reference materials, specimens, and/or chemicals, and make sure that the question lends itself to scientific inquiry. Ask specific, focused questions instead of broad, unanswerable questions. Questions might ask "how" something responds, forms, influences, or behaves, or how it is similar or different to something else.

Predict the answer to your question, and be prepared to accept the fact that your prediction might be incorrect or only partially correct. Examine and record all evidence gathered during testing that both confirms and contradicts your prediction.

Design a Testing Procedure that gathers information that can be used to answer your question. Make sure your procedure results in empirical, or measurable, evidence. Don't forget to do the following:

> Determine where and how the tests will take place – in a natural (field work) or controlled (lab) setting.

> Collect and use tools to gather information and enhance observations. Make accurate measurements. Use calculators and computers when appropriate.

> Plan how to document the test procedure and how to communicate and display resulting data.

> Identify variables, or things that might prevent the experiment from being "fair." Before beginning, determine which variables have no effect, a slight effect, or a major effect on your experiment. Create a method for controlling these variables.

Conduct the Experiment carefully and record your findings.

Analyze the Question Again. Determine if the evidence obtained and the scientific explanations of the evidence are reasonable based upon what is known, what you have learned, and what scientists and specialists have reported.

Communicate Findings so that others can duplicate the experiment. Include all pertinent research, measurements, observations, controls, variables, graphs, tables, charts, and diagrams. Discuss observations and results with relevant people.

Reanalyze the Problem and if needed, redefine the problem and retest. Or, try to apply what was learned to similar problems and situations.

Ongoing Projects: Problem Solving and Inquiry Scenarios

In the *Graphic Pages*, following the *Investigative Loop,* you will find the *Problem Solving and Inquiry Scenarios*. Photocopy this page for your students. Allow the students to work on one or more of these scenarios while completing this study of Insects and Arachnids. Although designed for intermediate students, the Problem Solving and Inquiry Scenarios are beneficial for all students' participation, if possible.

Experiences, Investigations, and Research

At the end of each lesson in the *Teacher's Section* is a category of activities entitled *Experiences, Investigations, and Research*. These activities expand upon concepts taught in the lesson, provide a foundation for further study of the content, or integrate the study with other disciplines. The following icons are used to identify the type of each activity.

Insects   Hands On   Geography   History   Writing   Websites   Literature   Math   Research

Cumulative Project

At the end of the program we recommend that students compile a Cumulative Project using the activities they have completed during their course of study. It may include the *Investigative Loops*, Lab Book, and the *3D Graphic Organizers* on display.

Please do not overlook the Cumulative Project, as it provides immeasurable benefits for your students. Students will review all the content as they create the project. Each student will organize the material in his or her unique way, providing an opportunity for authentic assessment and for reinforcing the context in which it was learned. This project creates a format where students can make sense of the whole study in a way that cannot be accomplished otherwise.

These 3D Graphic Organizers are used throughout Great Science Adventures.

Fast Food and Fast Folds

"If making the manipulatives takes up too much of your instructional time, they are not worth doing. They have to be made quickly, and they can be, if the students know exactly what is expected of them. Hamburgers, Hot Dogs, Tacos, and Shutter-folds can be produced by students, who in turn use these folds to make organizers and manipulatives." Dinah Zike

Every fold has two parts. The outside edge formed by a fold is called the **"mountain."** The inside of this edge is the **"valley."**

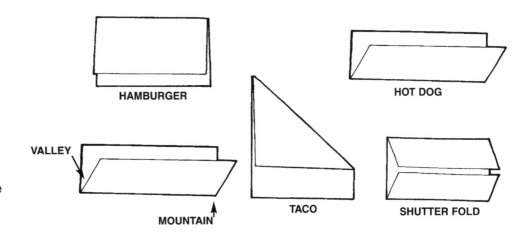

Storage – Book Bags

One-gallon reclosable plastic bags are ideal for storing ongoing projects and books that students are writing and researching.

Use a strip of clear, 2" tape to secure 1" x 1" pieces of index card to the top corner of a bag under the closure, front and back. Punch a hole through the index cards and the bag. Use a giant notebook ring to keep several of the "Book Bags" together.

Label the bags by writing on them with a permanent marker.

By putting the 2" clear tape along the side of the storage bag, and punching 3 holes in the tape, the bags can be kept in a notebook.

Half Book

1. Fold a sheet of paper in half like a Hamburger.

Large Question and Answer Book

1. Fold a sheet of paper in half like a Hamburger. Fold it in half again like a Hamburger. Cut up the valley of the inside fold, forming two tabs.

2. A larger book can be made by gluing Large Question and Answer Books "side by side."

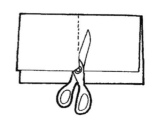

3 Tab Book

1. Fold a sheet of paper in half like a Hamburger or Hot Dog. Fold it into thirds. Cut up the inside folds to form three tabs.

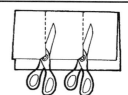

Pocket Book

1. Fold a sheet of paper in half like a Hamburger.

2. Open the folded paper and fold one of the long sides up two and a half inches to form a pocket. Refold along the Hamburger fold so that the newly formed pockets are on the inside.

3. Glue the outer edges of the side fold with a small amount of glue.

4. Make a multi-paged booklet by gluing several Pocket Books "side-by-side."

5. Glue a construction paper cover around the multi-page pocket booklet.

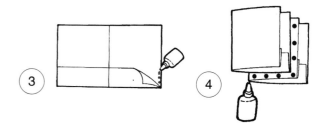

Vocabulary Book

1. Take two sheets of paper and fold each sheet like a Hot Dog.

2. Fold each Hot Dog in half like a Hamburger. Fold the Hamburger in half two more times and crease well. Unfold the sheets of paper, which are now divided into sixteenths.

3. On one side only, cut up the folds to the mountain top, forming eight tabs. Repeat this process on the second sheet of paper.

4. Take a sheet of construction paper and fold like a Hot Dog. Glue the solid back side of one vocabulary sheet, to one of the inside sections of the construction paper. Glue the second vocabulary sheet to the other side of the construction paper fold. (This step can be eliminated to form a one sided vocabulary book.) Make sure the center folds of the vocabulary books meet at the center fold of the construction paper.

5. Vocabulary Books can be made larger by gluing them "side-by-side."

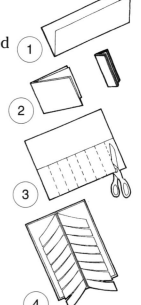

Layered Look Book

1. Stack two sheets of paper and place the back sheet one inch higher than the front sheet.

2. Bring the bottom of both sheets upward and align the edges so that all of the layers or tabs are the same distance apart.

3. When all tabs are an equal distance apart, fold the papers and crease well.

4. Open the papers and glue them together along the valley/center fold.

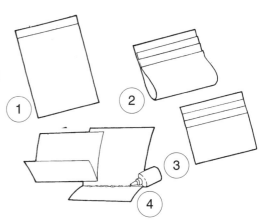

Trifold Book

1. Fold a sheet of paper into thirds.

2. Use this book as is, or cut into shapes. If the trifold is cut, leave plenty of fold on both sides of the designed shape, so the book will open and close in three sections.

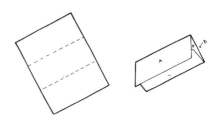

4 Door Book

1. Fold a sheet of paper into a Shutter Fold.

2. Fold it into a Hamburger.

3. Open the Hamburger and cut the Valley folds on the Shuttles only creating four tabs.

   Refold it into a Hamburger, with the fold at the top. Decorate the top sheet as the cover.

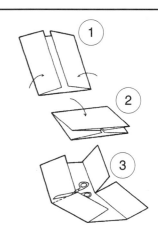

Side-by-Side

Some books can easily grow into larger books by gluing them side-by-side. Make two or more of these books. Be sure the books are closed, then glue the back cover of one book to the front cover of the next book. Continue in this manner, making the book as large as needed. Glue a cover over the whole book.

Small Question and Answer Book

1. Fold a sheet of paper in half like a Hot Dog.

2. Fold this long rectangle in half like a Hamburger.

3. Fold both ends back to touch the Mountain Top.

4. On the side forming two Valleys and one Mountain Top, make vertical cuts through one thickness of paper, forming tabs for questions and answers. These four tabs can also be cut in half making eight tabs.

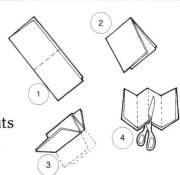

Bound Book

1. Take two sheets of paper and fold each like a Hamburger.

2. Mark both folds 1" from the outer edges.

3. On one of the folded sheets, "cut up" from the top and bottom edge to the marked spot on both sides.

4. On the second folded sheet, start at one of the marked spots and "cut out" the fold between the two marks. Do not cut into the fold too deeply; just shave it off.

5. Take the "cut up" sheet and roll it. Place it through the "cut out" sheet and then open it up. Fold the bound pages in half to form a book.

Variation...
To make a larger book, use additional sheets of paper, marking each sheet as explained in #3. Use an equal number of sheets for the "cut up" and "cut out." Place them one on top of the other and follow the directions in #4 and #5.

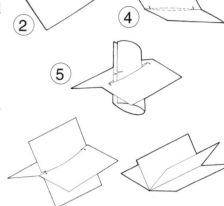

Pyramid Project

1. Fold a sheet of paper into a Taco.
   Cut off the excess tab formed by the fold.

2. Open the folded taco and refold it the opposite way, forming another taco and an X fold pattern.

3. Cut up one of the folds to the center of the X and stop. This forms two triangular-shaped flaps.

4. Glue one of the flaps under the other flap, forming a pyramid.

5. Set the Pyramid up on one end or glue two or more together to make a diorama.

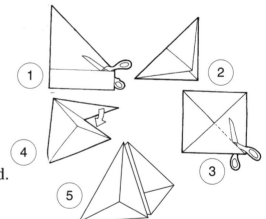

Display Box

1. Using a sheet of 8 1/2" x 11" paper, make a taco fold. Cut off the rectangle to form a square.

2. Fold the square into a shutter fold. Unfold and fold into another shutter fold, perpendicular to the previous one.

3. Cut along two fold lines on opposite sides of the large square. See diagram.

4. Collapse in and glue the cut tabs to form an open box.

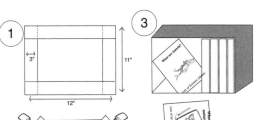

The *Lots of Science Library Book* Shelf

Make a bookshelf for the *Lots of Science Library Books* by using a box of an appropriate size or by following the instructions below.

1. Begin with an 11" x 12" piece of poster board or cardboard. Mark lines 3" from the edge of each side. Fold up along each line. Cut on the dotted lines as indicated in Illustration #1. Refold on the lines.

2. Glue the tabs under the top and bottom sections of the shelf. See Illustration 2. Cover your shelf with attractive paper.

3. If you are photocopying your *Lots of Science Library Books*, consider using green paper for the covers and the same green paper to cover your bookshelf.

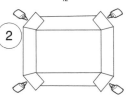

Great Science Adventures

Teacher's Section

Website addresses used as resources in this book are accurate and relevant at the time of publication. Due to the changing nature of the Internet, we encourage teachers to preview the websites prior to assigning them to students.

The authors and the publisher have made every reasonable effort to ensure that the experiments and activities in this book are safe when performed according to the book's instructions. We further recommend that students undertake these activities and experiments under the supervision of a teacher, parent, and/or guardian.

Note: Many of the activities require looking at insects. Insects are fun to catch and observe in their natural habitat, but be cautious when observing insects. Some insects sting or bite, and some people have severe allergic reactions. When you catch insects, treat them with care and return them to the site where you found them when possible. Wash your hands after handling insects.

If you would rather purchase insects, you may contact any of the following businesses. Some restrictions may apply.

1. Local pet stores, garden centers, and bait stores.
2. Carolina Biological Supply Company, 800–334–5551; 800–547–1733; http://www.carolina.com/
   Insects available: ants, crickets (nymphs and adults), ladybugs, mealworm larvae, milkweed bugs (eggs and adults), mosquitoes, praying mantis egg cases, silkworm eggs
3. Nasco Science, 800–558–9595; 414–563–2446
   Insects available: ants, crickets, ladybugs, mealworms (larvae, adults, pupae), painted lady butterflies, praying mantis egg cases
4. Ward's Biology, 800–962–2660
   Insects available: ants, ladybugs, mealworms (adult, larvae, eggs, pupae), praying mantis egg cases, silkworm eggs
5. The Biology Store, 800–654–0792
   Insects available: crickets, mealworms (larvae, pupae, adults), mosquito larvae
6. Connecticut Valley Biological Supply Company, 800–628–7748
   Insects available: ants, crickets (adults and nymphs), ladybugs, milkweed bugs (adult), mealworms (larvae, adults), painted lady butterflies, other butterfly species

Concept Map
Lessons 1, 2, & 21
Numbers Refer to Lesson Numbers

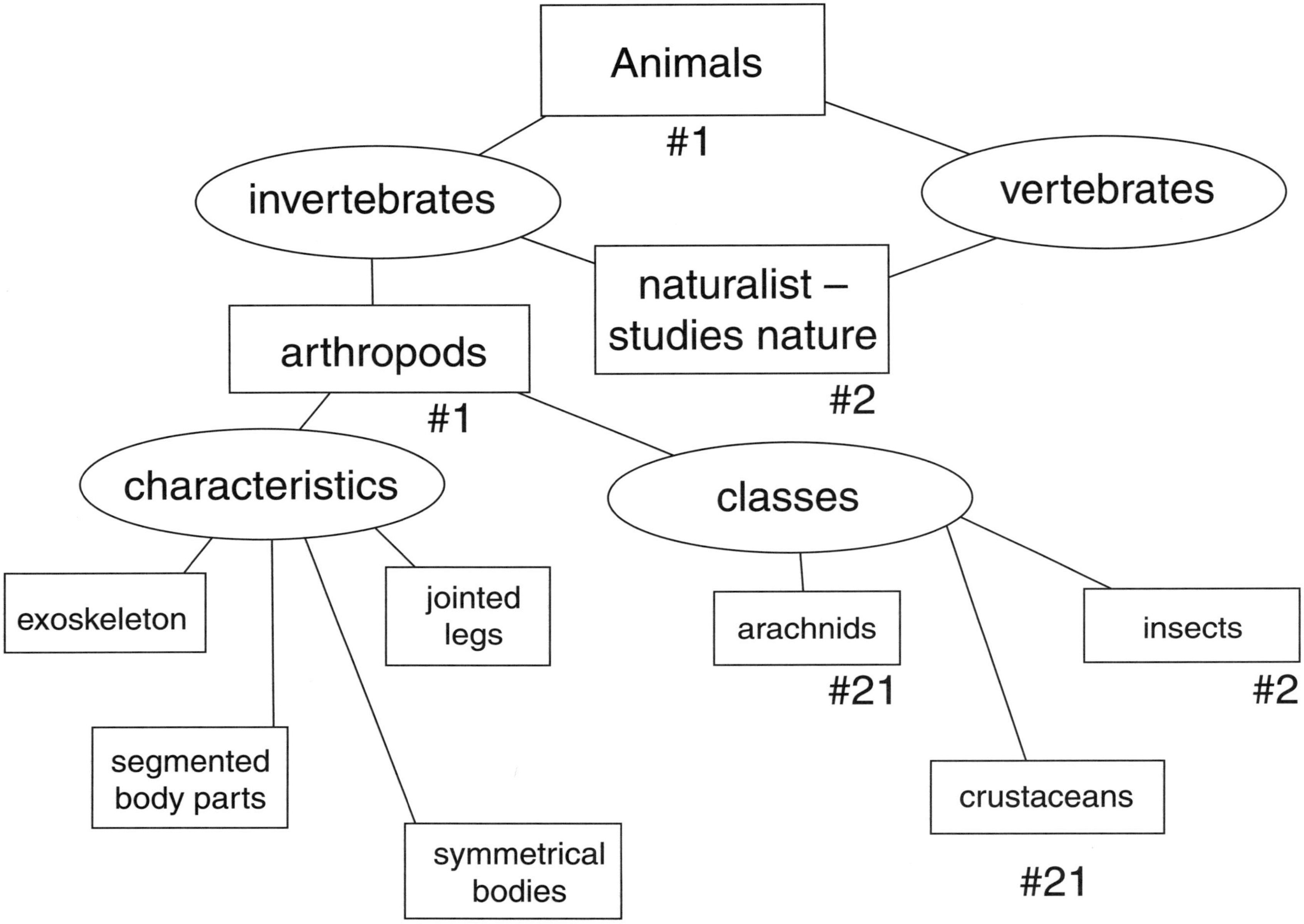

- Animals #1
  - vertebrates
  - invertebrates
    - arthropods #1
      - naturalist – studies nature #2
      - classes
        - insects #2
        - crustaceans #21
        - arachnids #21
      - characteristics
        - jointed legs
        - symmetrical bodies
        - segmented body parts
        - exoskeleton

Great Science Adventures

Lesson 1

What are insects?

Insect Concepts:
- Animals can be divided into two groups: vertebrates and invertebrates. Vertebrates have backbones; invertebrates do not.
- Animals are further classified into phylum, class, order, family, genus, and species.
- Arthropods make up the largest phylum.
- Insects make up the largest class of arthropods.
- Arthropods have an exoskeleton, segmented body parts, jointed legs, and symmetrical bodies.
- Insects have survived throughout time because they are small, they eat almost anything, and they can live almost anywhere.
- Prehistoric insects have been found preserved in hardened resin called amber.

Vocabulary Words: nature   classification   arthropod   insect
*symmetry   *vertebrates   *invertebrates   *taxonomy

Read: *Lots of Science Library Book #1.*

Classification
Kingdom
Phylum
Class
Order
Family
Genus
Species

Classification – Graphic Organizer

Focus Skill: classifying
Paper Handouts: 2 sheets of 8.5" x 11" paper   a copy of Graphic 1A
Graphic Organizer: Stack two sheets of paper together and make a Hot Dog. Cut along the fold. Stack the four sheets of paper together and make a Layered Look Book with eight layers and 1" tabs. Cut out the words in Graphic 1A. Glue them on the Layered Look Book beginning on the top tab: *Classification, Kingdom, Phylum, Class, Order, Family, Genus,* and *Species.*

✎ ✎✎ On the top and bottom section of the Kingdom tab, sketch the country where you live. On the bottom section of the Phylum tab, sketch the state (province or district) where you live. On the bottom section of the Class tab, write the name of your city. On the bottom section of the Order tab, write your street name. On the bottom section of the Family tab, write your house number. On the bottom section of the Genus tab, write your surname. On the bottom section of the Species tab, write your first name. Review how and why scientists classify living things.

✎✎✎ Complete ✎. Include examples for each classification.

Note: For post office box and rural route addresses, replace street signpost with P.O. Box or R.R.; replace house number with the number of post office box or number of rural route.

Insect Trap

Focus Skills: observation and identification
Activity Materials: glass jar   hand shovel   flat stone or brick
   four small rocks
Paper Handouts: 8.5" x 11" sheet of paper   a copy of Graphics 1B–C
Activity: Dig a hole in the ground large enough to hold the glass jar. Place the jar in the ground. Be sure the rim of the jar is level with the top of the hole. Place some leaves and grass in the jar. Place four rocks around the rim of the jar and put the flat stone on the four small rocks. Check the jar every few hours. Do not keep insects captive for more than 24 hours. Release them at the location where they were collected, when possible.
Graphic Organizer: Fold Graphic 1B into a Shutter Fold. Cut on the dotted lines. Cut out Graphic 1C and glue the pictures on the front of the Shutter Fold. Open the Shutter Fold.
✎ Draw pictures of insects you found in your insect trap.
✎✎ Complete ✎. Identify the insects and label them.
✎✎✎ For each insect found in the trap, sketch it and record the following information: time of day it was found and qualitative description.

Symmetry

Paper Handouts: 8.5" x 11" sheet of paper   a copy of Graphic 1D
Activity: An insect has a symmetrical design. One side is a mirror image of the other. Fold on the center line and cut through both sides of Graphic 1D. Now, open it. Is one side of the fold the same as the other side? Look around the room and find things that are symmetrical.

Make a Pond

Activity Materials: piece of plastic pond lining (size depends on the pond size you determine)
   pond plants (available at aquarium/pet stores or garden stores)
Activity: Ponds are a perfect habitat for many insects. Many man–made ponds are elaborate, but you can make a simple one. Discuss the best place to make your pond. Dig a hole about 18 inches deep. Place the sheet of plastic in the hole so that the plastic overlaps the edges of the hole. Place a little soil in the bottom of the hole. Place heavy rocks or bricks around the perimeter of the hole to keep the lining in place. Fill with water. Add pond plants and pondweed. Observe your pond throughout the time you study insects.

Optional: For an even simpler pond, just place a plastic tub outside and fill it with water. Collect water plants from a nearby pond (or purchase them from an aquarium/pet store) and put them in the water. Observe your pond.

Experiences, Investigations, and Research

Select one or more of the following activities for individual or group enrichment projects. Allow your students to determine the format in which they would like to report, share, or graphically present what they have discovered. This should be a creative investigation that utilizes your students' strengths.

 1. Fold a sheet of paper in half. Open the paper and place two dabs of poster paint in the middle of the fold. Fold the paper carefully and smooth the crease. Open the sheet of paper and describe the symmetrical design. Investigate and describe a butterfly's wings in the same manner.

 2. Research how scientists use amber and impression fossils to understand prehistoric insects.

 3. *Who, What, When, Where:* Research Carl Linnaeus, the father of taxonomy. Make a 4 Door Book to report your data.

 4. Visit a natural history museum and view its insect collection.

 5. Use a nature guidebook or real insects to observe their symmetrical design.

 6. Read: *Carl Linnaeus: Father of Classification* by Margaret Jean Anderson

 7. http://sln.fi.edu/tfi/units/life/classify/classify.html (Franklin Institute Online)

 8. http://library.thinkquest.org/11771/english/hi/biology/taxonomy.shtml (The Sciences Explorer)

 9. http://www.discovery.com/area/science/micro/cockroach.html

Notes

Great Science Adventures

Lesson 2

Who studies insects?

Insect Concepts:
- Naturalists are people who study and observe nature.
- People who specialize in the study of insects are called entomologists.
- Naturalists and entomologists observe, listen, record, and sketch what they see.
- Insect research can be conducted in a natural setting or in a controlled lab environment.

Vocabulary Words: collection   naturalist   *identification   *entomology

Read: Lots of Science Library Book #2.

Teacher's Note: Choose one type of Insect Collection for the students to work on during this program. Two types of insect collections are made of paper found in the Graphics Pages. The paper insects can be stored in boxes or in a book. These types of insect collections are described below in Options 1 and 2. The other type of insect collection involves finding or ordering real insects. This is explained below in Option 3.

Insect Collection – Option 1

Paper Handouts: 9 sheets of colored construction paper.
Graphic Organizer: Make nine Display Box Organizers. See page 5. Glue them together side by side as shown. Directions for the paper insects are located in various lessons. You will be adding to this collection through Lesson 19. This will be referred to as the *Paper Insect Collection*. A separate Display Box Organizer will be made for Arachnids in Lesson 22.

Insect Collection – Option 2

Paper Handouts: 4 sheets of 8.5" x 11" paper   a copy of Graphic 2A
Graphic Organizer: Make a Bound Book out of the four sheets of paper. Cut out Graphic 2A and glue it on the cover of the Bound Book. Title it *Insects and Arachnids*. You will add paper insects to this through Lesson 19. Later, in Lessons 22–24, you will use this same Bound Book to add arachnids. This will be referred to as the *Insect and Arachnid Bound Book*.

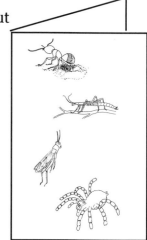

Insect Collection – Option 3

Activity Materials: display board or several foam meat trays   long straight pins
Paper Handouts: 5 sheets of 8.5" x 11" paper
Graphics Organizer: Fold each sheet of paper into a Hamburger. Cut on the folds. Make Display Boxes with each half sheet of paper. Glue nine of them together as shown. Cut pieces of display board to fit inside each Display Box. This display is for mounting real insects. You will add insects through Lesson 19. Directions are found in the lessons. Later in Lessons 22–24, you will add arachnids to your collection. This will be referred to as the *True Insect and Arachnid Collection*.

Use the following guidelines when mounting insects.
1) Pin beetles through right wing cover.
2) Pin true bugs through scutellum*.
3) Pin flies slightly to the left of the thorax.
4) Pin bees, wasps, and ants through the middle of the thorax.
5) Pin grasshoppers, dragonflies, butterflies, and moths through the thorax.
6) Pin plant bugs, cicadas, water skaters, and assassin bugs through the scutellum.
7) To mount very small insects, cut small triangles of cardboard. Place a dot of glue on the point of the triangle and glue the insect to this point. The triangle can then be pinned to a mount board.

   * Note: The scutellum is a small triangle found between the wings and just behind the thorax.

Catching Flying Insects

Focus Skill: identifying
Paper Handouts: 8.5" x 11" sheet of paper   a copy of Graphic 2B
Activity Materials: clear plastic cup or jar   magnifying glass
   piece of cardboard a little larger than the mouth of the jar
Activity: The best way to catch flying insects is to let them land on a leaf and then place the cup upside down over it. The insect will fly upward, so quickly place the piece of cardboard under the cup and pull away from the leaf. Observe the insect with a magnifying glass.
Graphic Organizer: Make a Trifold Book. Glue Graphic 2B on the cover. Color the butterfly. Open the Trifold Book.
✎   Draw pictures of the flying insects you caught or observed.
✎✎   Complete ✎. Identify the insects and label the sketches.
✎✎✎   Choose three insects and on each section of the Trifold Book (top, middle, bottom), sketch one insect and record the following information: location where insect was found and plants in the area.

Insect Net

Activity Materials: 5–gallon nylon paint strainer (affordable and available at paint stores)
   wire coat hanger   36" wooden dowel   duct tape
Activity: Form the coat hanger into a square. Sew the paint strainer onto the square coat hanger. Tape the coat hanger net to the wooden dowel. To catch insects, use a smooth sweeping motion back and forth. When an insect is caught, quickly flick your wrist. See page 62.

Experiences, Investigations, and Research

Select one or more of the following activities for individual or group enrichment projects. Allow your students to determine the format in which they would like to report, share, or graphically present what they have discovered. This should be a creative investigation that utilizes your students' strengths.

 1. Discover where insects live and why. Look indoors where food, especially grain, is stored. Look in more secluded areas, such as closets and basements. Look outdoors in a variety of places such as a manicured yard, a vacant lot, and the woods. Look on plants and under logs, rocks, and bark. Did you find more insects inside or outside? where traffic was heavy or light? where plants were manicured or wild? Why? Where did you find the least number of insects? Why? Did you find live insects? Did you find dead insects?

 2. *Who, What, When, Where*: Research the life of naturalists Jean Henri Fabre or Thomas Say.

 3. Research insects and their impact throughout history. Example: bubonic plague and locust-induced famines.

 4. Investigate the pros and cons of collecting insect specimens.

 5. Research insects that are endangered or extinct. Explain the causes and effects of insect extinction.

 6. Read *Children of Summer: Henri Fabre's Insects* by Margaret Jean Anderson

 7. http://www.bartleby.com/65/fa/Fabre–Je.html (Bartleby.com Great Books Online)

 8. http://www.museum.unl.edu/research/entomology/index.htm (University of Nebraska State Museum)

Insects Concept Map
Lessons 3-7
Numbers Refer to Lesson Numbers

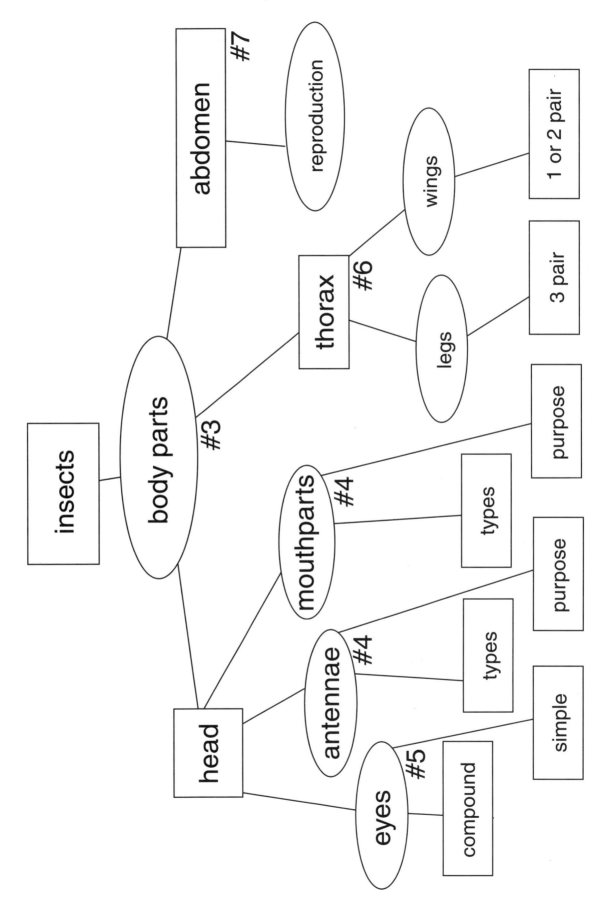

Great Science Adventures

Lesson 3

What are the characteristics of insects?

Insect Concepts:
- Insects do not have internal skeletons, but they are covered with a protective exoskeleton.
- An exoskeleton does not grow with an insects. Insects molt, or shed, their exoskeletons numerous times as their bodies change sizes, or develop.
- An insect's body has thee main segments or parts: head, thorax, and abdomen.
- Specialized eyes, mouthparts, and antennae are found on an insect's head.
- An insect's three pairs of legs are attached to its thorax.
- Insect legs are adapted for such tasks as grasping, jumping, digging, swimming, and collecting.
- Most adult insects have one or two pairs of wings attached to the thorax.
- A few species of insects have wings only during certain developmental stages, in some species, only one sex has wings.
- An insect's abdomen houses organs used for digestion, elimination, and reproduction.
- The females of some insect species have an ovipositor at the end of their abdomen for laying eggs.

Vocabulary: exoskeleton   molts   head   thorax   abdomen   *chitin (KYE tin)

Read: *Lots of Science Library Book #3.*

All About Insects – Graphic Organizer
Paper Handouts: 12" x 18" construction paper   a copy of Graphics 3A–C
Graphic Organizer: Fold the construction paper into a Hot Dog. Glue Graphics 3A–C on the front. Cut the Hot Dog to make a 3 Tab Book. Throughout the next four lessons, you will be adding to this Graphic Organizer. You may fold the Graphic Organizer at the cuts for easy storage. It will be referred to as the *All About Insects Graphic Organizer.*

Paper Insect Collection
Focus Skills: graphing, identifying members of a group
Paper Handouts: a copy of Graphics 3D–E
Graphic Organizer: Cut out Graphic 3D. Color it and glue it together. Place the insect in one of the boxes of your *Paper Insect Collection*. Cut out Graphic 3E, the Insect Data Card. Read the data and graph the number of species. Place the Insect Data Card in the same box as the insect.

Insect and Arachnid Bound Book

Focus Skills: graphing, identifying members of a group
Paper Handouts: a copy of Graphics 3D–E
   Cut out Graphic 3D. Cut out a dorsal, or top, view and a lateral, or side, view of the insect. Color them and glue them on the first page of the *Insect and Arachnid Bound Book*. Cut out Graphic 3E, the Insect Data Card. Read the data, graph the number of species, and glue beneath the pictures of the insect.

True Insect and Arachnid Collection

Focus Skills: graphing, identifying members of a group
Paper Handouts: a copy of Graphic 3E
   Add a common housefly to your *True Insect and Arachnid Collection*. Cut out Graphic 3E. Read the Insect Data Card, graph the number of species, and pin below your insect. If you cannot find the specified insect, you may mount any insect of your choice. Make your own Insect Data Card.

Insect Body Parts

Activity Materials: egg carton   4 black pipe cleaners
Activity: Cut out three sections of the egg carton in one piece. Punch holes and put the pipe cleaners through the middle carton (thorax) for legs. Bend the legs at the joints. Put another pipe cleaner in the head for antennae, make a V, and twist. With felt marker, draw the eyes and mouth. Review the body parts of an insect.

Experiences, Investigations, and Research

Select one or more of the following activities for individual or group enrichment projects. Allow your students to determine the format in which they would like to report, share, or graphically present what they have discovered. This should be a creative investigation that utilizes your students' strengths.

 1. Trees are hosts to many different kinds of insects on their trunk, leaves, branches, flowers, fruits, and seeds. Place an old, light–colored sheet on the ground under a tree. Grasp a branch and shake it gently but firmly. With tweezers, pick up the insects and put them in a jar. Look at them with a magnifying glass.

 2. Look in magazines to find pictures of insects, mammals, reptiles, and fish. Compare an insect to other animals. How are they the same? How are they different?

 3. Play *Simon the Insect Says*. Pretend you are an insect. Your chest is the thorax and your stomach is the abdomen. Head, legs, and mouth are the same. Choose one person to be Simon the Insect. He begins by saying "Simon the Insect says, 'Touch your thorax." Continue play.

 4. Many of our foods contain insects. Research the laws that govern standards for the Food Defect Action Levels (FDAL) set by the Department of Health and Human Services to determine how many insect parts can legally be in processed foods.

 5. http://gnv.ifas.ufl.edu/~tjw/recbk.htm

 6. http://www.theaes.org/ (Amateur Entomologists' Society)

Great Science Adventures

Lesson 4

How do insects use and their mouths and antennae?

Insect Concepts:
- Insects' antennae are most frequently used to smell and feel.
- Most insects have two segmented antennae on their heads, located between their eyes.
- Antennae have developed into many different shapes, including feathery, twisted, and cone-shaped.
- Some insects secrete chemical signals called pheromones that are sensed by the antennae of insects of the same species.
- Insects have two main types of mouthparts – mandibles and maxillae. These have adapted individually into many variations, and some orders of insects have combinations of both.
- Mouthparts may be used for piercing, chewing, siphoning, holding, and tearing. Wood-eaters make up the largest group of insects.

Vocabulary: antennae   smell   taste   touch   *olfactory nerves (ol FAK to ree)   *mandibles   *maxillae (max IL ee)   *labium (LAY bee um)

Read: *Lots of Science Library Book #4.*

Insect Mouth and Antennae – Graphic Organizer

Focus Skill: explaining functions of parts
Paper Handouts: 2 pieces of paper, 4.25" x 5.5" (This is equivalent to 1/4 of an 8.5" x 11" sheet of paper.)
a copy of Graphics 4A–B   *All About Insects Graphic Organizer*
Graphic Organizer: Make Matchbooks out of the two pieces of paper. Fold the bottom tab up 1/2 inch. Cut out and color Graphics 4A–B. Glue one on the front of each Matchbook. Label the tabs *Antennae* and *Mouth* accordingly. Open the Antennae Matchbook.

- ✎ Draw pictures of one or more antennae types.
- ✎✎ On the top section, complete ✎. On the bottom section, write clue words about the uses of antennae: *smell, taste, and touch.*
- ✎✎✎ Open the Matchbook and on the top section, sketch a pair of antennae showing the three parts and label them. On the bottom section, explain the uses of antennae, using your vocabulary words.

  Open the Mouth Matchbook.
- ✎ Draw a picture of an insect's mouth.
- ✎✎ On the top section, complete ✎. On the bottom section, write clue words about insects' mouths: *pierce, chew, sponge.*
- ✎✎✎ On the front of the Matchbook, label the mouthparts. Open the Matchbook and on the top section, describe the different mouthparts and their functions. On the bottom section, include an example of an insect with each type of mouth.

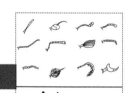
Mouth

Glue the Antennae Matchbook and the Mouth Matchbook in your *All About Insects Graphic Organizer* under the Head tab, on the top section.

Different Types of Mouths

Activity: Think about the different types of mouths insects have: chewing, piercing, sucking, and sponging. Then think of common objects that duplicate the action. For example, a needle pierces and a straw sucks. Now, think of insects that have the various mouthparts. For example, a mosquito has a piercing mouth and a housefly has a sponging mouth.

Experiences, Investigations, and Research

Select one or more of the following activities for individual or group enrichment projects. Allow your students to determine the format in which they would like to report, share, or graphically present what they have discovered. This should be a creative investigation that utilizes your students' strengths.

 1. Insects rely on their antennae to sense their surroundings. Blindfold a partner and give him or her a walking stick. Take a walk together. Watch your partner and protect him or her from danger. How well did your partner get around? Trade places with your partner. Relate this experience to an insect's use of its antennae.

 2. Observe ants and how they use their antennae.

 3. Look in the mirror. Open your mouth and examine your teeth. Do you have special teeth with different functions? biting teeth? chewing teeth? Why do you use a straw to drink liquids?

 4. List several types of insects that have specialized mouthparts. Next to each type of insect, describe what they eat.

 5. *Who, What, When, Where:* Sir Ronald Ross discovered that mosquitoes transmitted malaria. Research Ross and the importance of his discovery.

 6. Read and discuss *Two Bad Ants* by Chris Van Allsburg

 7. http://www.orkin.com/Main.htm

Great Science Adventures

Lesson 5

How do insects see?

Insect Concepts:
- Insects probably see very clearly up close, but have blurred sight at a distance.
- Insect eyes are always open because they do not have eyelids to cover them.
- Some insects can see light rays that are invisible to humans, such as ultraviolet and infrared.
- Most mature insects have two compound eyes that dominate the insect's head.
- Compound eyes are composed of a few thousand lenses, or hexagon–shaped facets.
- Often adult insects will also have three simple eyes, or ocelli, located between the compound eyes that help the insect differentiate between light and dark.

Vocabulary: vision   compound   eyes   lenses   *facets   *lateral   *dorsal   *ultraviolet   *ocelli (o SEL ee)   *ommatidia (o mah TEEd ee ah)

Read: *Lots of Science Library Book #5.*

Paper Insect Collection

Focus Skills: graphing, identifying members of a group
Paper Handouts: a copy of Graphics 5A–B
Graphic Organizer: Cut out Graphic 5A. Cut, color, and glue together. Place the insect in one of the boxes of your *Paper Insect Collection*. Cut out Graphic 5B, read the data, and graph the number of species. Place the Insect Data Card in the box with the insect.

Insect and Arachnid Bound Book

Focus Skills: graphing, identifying members of a group
Paper Handouts: a copy of Graphics 5A–B
Cut out Graphic 5A. Cut out the top, or dorsal, view and a lateral, or side, view of the insect. Color and glue on the next page of the *Insect and Arachnid Bound Book*. Cut out Graphic 5B, read the data, graph the number of species, and glue it on the bottom of the page.

True Insect and Arachnid Collection

Focus Skills: graphing, identifying members of a group
Paper Handouts: a copy of Graphic 5B
Add a dragonfly to your collection. Read the Insect Data Card, graph the number of species, and attach it near your insect. If you cannot find the specified insect, you may mount any insect of your choice. Make your own Insect Data Card.

Dragonfly
Class: Insect
Order: Odonata
Number of Species: 5,300
Characteristics: keen eyesight; small antennae; strong, biting mouthparts
Wings: long, thin membranous wings
Habitat: near ponds or lakes
Diet: carnivorous

300,000
200,000
100,000
50,000
Number of Species

Insect Eyes – Graphic Organizer

Focus Skill: explaining parts of a whole
Paper Handouts: a 4.25" x 5.5" piece of paper   a copy of Graphic 5C
All About Insects Graphic Organizer
Graphic Organizer: Make a Matchbook with a ½" tab. Glue Graphic 5C on the front of the Matchbook. Open the Matchbook.
- ✎ Draw a picture of insect eyes.
- ✎✎ Write clue words describing insect eyes: *compound eyes have many lenses, simple eyes have one lens.*
- ✎✎✎ Explain the parts of insect eyes, and describe how they see.

Glue the Eye Matchbook on the *All About Insects Graphic Organizer* under the Head tab on the bottom section.

Compound Eyes – Investigative Loop Lab 5–1

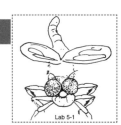

Focus Skill: observing
Lab Materials: 12 drinking straws cut in half   tape   scissors
Paper Handouts: 8.5" x 11" sheet of paper   a copy of Lab Graphic 5–1
     index cards
Graphic Organizer: Make a Pocket Book. This is the students' Lab Book. It will be used in this and future lessons. Glue Lab Graphic 5–1 on the left pocket. Store the Lab Record Cards in this pocket.
Question: What images do insects see with compound eyes?
Research: Review *Lots of Science Library Book #5* and review the Question.
Predictions: Predict what insects see with compound eyes. Write your prediction on a Lab Record Card.
Procedure: Divide the straws into two groups of 12 straw pieces each. Bundle each group and tape them together. Place them in front of your eyes like binoculars to see what an insect compound eyes see. Look at a simple object with your own eyes. Now, look at the same object with the bundled straws.
Observations: Explain how the simple object looked with your eyes. Explain how it looked through the bundled straws.
Record the Data: Label a Lab Record Card "Lab 5-1." Draw what you saw through the bundled straws.
Conclusions: Why do you think the image you saw through the bundled straws appeared as it did? Draw conclusions about how insects view the world, based on this lab.
Communicate the Conclusions: On a Lab Record Card, compare your observations and conclusions with your predictions. **Note: Scientists continue to study how insects see.**
Spark Questions: Discuss questions sparked by this lab.
New Loop: Choose one question to investigate further,
     OR make a New Loop, using the following information and the *Investigative Loop* found in the Graphics Pages:

Lab Materials: red cellophane   2 rubber bands
   2 sets of bundled straws from the activity above

Procedure: Cover the bundled straws with red cellophane. Secure them with rubber bands.

Observations: Look through the bundled straws and observe various colored items. Look at a red object.

Conclusions: What did you see through the red cellophane? Describe the appearance of red objects through the red cellophane. Draw conclusions about how insects view the world.

 Design Your Own Experiment: Select a topic based upon the experiences in the *Investigative Loop*. See page viii for more details.

Experiences, Investigations, and Research

Select one or more of the following activities for individual or group enrichment projects. Allow your students to determine the format in which they would like to report, share, or graphically present what they have discovered. This should be a creative investigation that utilizes your students' strengths.

1. Catch insects or find dead ones. Observe their eyes. Do they have compound eyes? Locate the facets. If they have simple eyes, how many do they have?

2. *Ants–on–a–Log / Aphids–on–a–Log* Slice celery into 2–inch pieces. Spread with peanut butter and sprinkle with raisins to represent ants. Prepare the celery as above and sprinkle with sesame seeds to represent aphids.

3. Insects are often attracted to light. Observe light fixtures or remove lampshades and empty the contents onto a white sheet of paper. What did you find?

4. Research the color spectrum that insects can see.

5. http://www.insect-world.com/ (Insect World – Britannica Internet Guide Award)

Notes

Great Science Adventures

Lesson 6

What is the form and function of insects' legs and wings?

Insect Concepts:
- Insects use their legs for walking, jumping, swimming, digging, catching prey, and collecting pollen.
- Insects are the only flying invertebrates.
- Most flying insects have two pairs of wings, but some have one and a few have none.
- When insects have two pairs of wings, the front wings usually protect the hind wings, which are often folded when not in use.
- The flapping rates of insect wings vary considerably.

Vocabulary: legs   wings   *coxa (KOX ah)   *femur (FEE mer)   *tibia (TIB ee ah)   *tarsus (TAR sus)

Read: *Lots of Science Library Book #6.*

Insect Legs and Wings – Graphic Organizer

Focus Skill: describing components
Paper Handouts: 2 pieces of paper, 4.25" x 5.5"   a copy of Graphics 6A–B
All About Insects Graphic Organizer
Graphic Organizer: Make two Matchbooks with ½" tabs. Glue Graphics 6A on the front of one Matchbook and 6B on the front of the other one. Label the tabs *Legs* and *Wings* accordingly. Open the Legs Matchbook.
- ✎ Draw a picture of insect legs.
- ✎✎ Complete ✎ on the top section. On the bottom section, write clue words about insect legs: *six legs, segmented, attached to thorax.*
- ✎✎✎ On the front of the Legs Matchbook, label the leg parts. Open the Legs Matchbook. On the bottom section, describe insect legs using your vocabulary words.

Open the Wings Matchbook.
- ✎ Draw a picture of insect wings.
- ✎✎ Complete ✎ on the top section. On the bottom section, write clue words about insect wings: *some insects have one or two pairs of wings attached to the thorax; some insects have no wings.*
- ✎✎✎ Open the Wings Matchbook. On the top section, explain how insects fly and describe insect wings. On the bottom section, include examples of insects with no wings, one pair of wings, and two pairs of wings.

Glue the Legs and Wings Matchbooks on your *All About Insects Graphic Organizer* under the Thorax tab on the top section.

Grasshopper Jump

Activity Materials: Chinese jump rope, or make one with a bag of rubber bands. Connect several rubber bands with slip knots.

Activity: A grasshopper can jump about two feet high. Ask two people to hold the rope two feet above the floor. If no one is available, tie the ends to chairs or other furniture. Be a grasshopper and jump the rope. Did you do it? For younger students, begin with a lower height and increase gradually. As a challenge for older students, continue to raise the jump level.

Note: If a human jumped the same height as a grasshopper in ratio to his height, he would have to jump over a tall building.

Insect Noises – Investigative Loop Lab 6-1

Focus Skill: experiencing a concept
Lab Materials: paper cup   wax paper   rubber band
Paper Handouts: Lab Book   a copy of Lab Graphic 6-1   Lab Record Cards
Graphic Organizer: Glue Lab Graphic 6-1 on the right pocket of the Lab Book.
Question: How do flying insect sounds compare with each other?
Research: Review *Lots of Science Library Book #6* and review the Question.
Predictions: Predict whether the sounds made by flapping insect wings will be different from each other.
Procedure: Catch several different kinds of flying insects. Place them individually in paper cups. Cover the cups with wax paper and secure them with rubber bands.
Observations: Put each cup up to your ear and listen.
Record the Data: Label a Lab Record Card for each insect with "Lab 6-1." On each card, draw the insect and describe the sound it makes.
Conclusions: Review the Lab Record Cards. Did all the insects make the same sound? How did they differ? How were they the same?
Communicate the Conclusions: On a Lab Record Card, compare your observations and conclusions with your predictions. Share your Lab Record Cards with one person who did not participate in the activity.
Spark Questions: Discuss questions sparked by this activity.
New Loop: Choose one question and investigate it further.

Optional: Try imitating the sounds of the captured insects and have a partner guess the insect. Trade places with your partner.

✎✎✎ **Design Your Own Experiment:** Select a topic based upon the experiences in this *Investigative Loop*. See page viii for more details.

Clay Insects

Activity Materials: modeling clay or dough
Activity: Make insects with modeling clay or dough. Make them as detailed as possible.

Experiences, Investigations, and Research

Select one or more of the following activities for individual or group enrichment projects. Allow your students to determine the format in which they would like to report, share, or graphically present what they have discovered. This should be a creative investigation that utilizes your students' strengths.

 1. Catch flying insects or find dead ones. If they are alive, observe their wings as they fly about in a bug jar. Observe and illustrate the shape of their wings. How many wings do they have? Examine the color and veins.

 2. Catch insects or find dead ones. If they are alive, observe how they move about on their legs. What do they use their legs for? Jumping? Digging? Swimming?

 3. Find several grasshoppers or crickets, available through pet stores or bait shops. Place a yardstick on the ground and mark a start line. Predict how far the insect will jump. Hold the insect carefully and release. Measure the distance and record. Try it again, measure, and record. Do the same with another grasshopper. How close were your predictions? Did you see any marked differences among them? Graph your results.

 4. Find one or two partners and figure out a way to walk like an insect.

 5. http://www.insects.org/index.html

 6. http://www.biologists.org/ (The Company of Biologists Limited)

Notes

Great Science Adventures

Lesson 7

How do insects reproduce?

Insect Concepts:
- Adult insects' sole purpose is reproduction of the species.
- Most insects reproduce sexually when a female sex cell is united with a male sex cell.
- Insects use a variety of creative ways to attract mates, including smell, touch, sound, and special displays.
- Insects lay eggs that vary greatly in size, shape, and number.
- Some eggs are laid singly, while others are laid in groupings.
- Different species of insects lay eggs in the ground, on water, on other animals, in waste, and on plants.
- Insects lay eggs in areas that will provide food for the newly hatched larvae to consume.

Teacher's Note: An alternative assessment suggestion for this lesson is found on pages 82–83. If Graphic Pages are being consumed, first photocopy assessment graphics that are needed.

Vocabulary: reproduction   egg   hatch   courtship   mating   scraper   file
*tympanum (tim PAN um)   *fertilize   *sperm   *pheromones (FER o mohnz)

Read: *Lots of Science Library Book #7.*

Paper Insect Collection

Focus Skills: graphing, identifying members of a group
Paper Handouts: a copy of Graphics 7A–B
Graphic Organizer: Cut out Graphic 7A. Cut, color, and glue together. Place in one of the boxes of your *Paper Insect Collection.* Cut out Graphic 7B, read the data, and graph the number of species. Place the Insect Data Card in the box.

Insect and Arachnid Bound Book

Focus Skills: graphing, identifying members of a group
Paper Handouts: a copy of Graphics 7A–B
Graphic Organizer: Cut out Graphic 7A. Cut out a dorsal view and a lateral view of the insect. Color and glue on the next page of the *Insect and Arachnid Bound Book.* Cut out Graphic 7B, read the data, graph the number of species, and glue beneath the picture of the insect.

True Insect and Arachnid Collection

Focus Skills: graphing, identifying members of a group
Paper Handouts: a copy of Graphic 7B
    Add a cockroach to your collection. Read the Insect Data Card, graph the number of species, and pin below your insect. If you cannot find the specified insect, you may mount any insect of your choice. Make your own Insect Data Card.

Insect Reproduction – Graphic Organizer

Reproduction

Focus Skill: explaining a process
Paper Handouts: a piece of paper, 4.25" x 5.5"   a copy of Graphics 7C
  All About Insects Graphic Organizer
Graphic Organizer: Make a Matchbook with a ½" tab. Glue Graphic 7C on the front of the Matchbook. Label the tab *Reproduction*. Open the Matchbook.
  ✎ Draw different types of insect eggs (small, oval, capsules).
  ✎✎ Write clue words explaining insect reproduction: *most insects are sexual, insects lay eggs*.
  ✎✎✎ Using your vocabulary words, explain various ways insects attract mates and reproduce.

Glue the Reproduction Matchbook in the *All About Insects Graphic Organizer* under the Abdomen tab on the top section.

Cricket's Call

Activity Materials: piece of paper   nail file
Activity: The edge of the paper represents the cricket's scraper, or bumpy ridge on the cricket's wing. The nail file represents its file, or a sharp edge of its wing. Rub the file against the edge of the paper. The sound is similar to crickets chirping.

Cricket Weatherman

Activity: Find out the temperature with a little help from a cricket. Listen to tree crickets. Count the number of chirps in 15 seconds. Add 40, and you have the Fahrenheit temperature. For example, if you hear 25 chirps, just add 40. 25 + 40 = 65. The temperature is 65° F. To determine the temperature in Celsius, count the number of chirps in 15 seconds. Divide this number by 2, then add 6. For example, if you hear 30 chirps, just divide by 2 to get 15, then add 6. The temperature is 21°C.

Tracking a Scent – Investigative Loop Lab 7–1

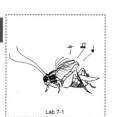

Focus Skill: experiencing a concept
Lab Materials: several small containers   cotton balls   vanilla extract
  vinegar   root beer   lemon
  (You may choose other scents.)
Paper Handouts: 8.5" x 11" sheet of paper   Lab Book   Lab Record Cards
  a copy of Lab Graphic 7–1
Graphic Organizer: Make a Pocket Book. Glue it side–by–side to the Lab Book. Glue Lab Graphic 7–1 to the left pocket.
Concept: Insects use scent to locate objects.
Research: Read *Lots of Science Library Book #7* and review insects' use of smell.
Procedure: Prepare two cotton balls for each scent and place in containers. Place the containers around the room. Give your partner one of the cotton balls to smell and ask him to find its mate, the matching scent. Try this again outside. Trade places with your partner.
Observations: Describe the process of locating the scent inside. Describe the same process outside.
Record the Data: Label two Lab Record Cards "Lab 7–1." On one card describe the process of locating the scent inside. On the other card describe the process outside.
Conclusions: Review the Lab Record Cards and compare the two procedures. How did they differ? How were they the same? Where do most insects use their senses to locate objects, inside or outside? Draw conclusions about insects' ability to locate with smell from this lab.

Communicate the Conclusions: Label a Lab Record Card "Lab 7–1." Write the conclusions on the card or write it in a newspaper format.

Spark Questions: Discuss questions sparked by this lab.

New Loop: Investigate one of the questions further.

✎✎✎ **Design Your Own Experiment:** Select a topic based upon the experiences in the *Investigative Loop*. See page viii for more details.

Galls

Note: Complete this activity in late summer or fall when galls are abundant.

Activity Materials: glass jar   small square of gauze or light cotton   rubber band

Activity: Look for galls on oak trees or other plants. Collect the gall by breaking off a twig or stem of a plant. Place it in the jar, cover the jar opening with gauze, and secure it with a rubber band. Place the container in a safe place fairly consistent with the outdoor temperature, such as in the garage or on a porch. You may want to collect several galls. Place each gall in a separate container with a piece of twig or stem. Label each one with the name of the plant. Check the gall as spring approaches. **You should find that adult insects emerge from the galls.**

Experiences, Investigations, and Research

Select one or more of the following activities for individual or group enrichment projects. Allow your students to determine the format in which they would like to report, share, or graphically present what they have discovered. This should be a creative investigation that utilizes your students' strengths.

  1. Find a cricket. Locate its "ears," found on the front leg right below the knee. Look for a small swelling that is usually a pale color.

2. Research courtship behaviors of several insects.

3. Research the 17–year reproduction cycle of the cicada.

4. http://entomology.si.edu/ (Smithsonian Institute Department of Entomology)

Insects Concept Map
Lessons 8-9
Numbers Refer to Lesson Numbers

- insects
  - systems
    - respiratory #8
      - trachea
      - spiracles
    - circulatory #8
      - hemolymph
      - dorsal vessel
    - digestive #9
      - foregut
      - midgut
      - hindgut
    - nervous #9
      - brain
      - nerve cords

Great Science Adventures

Lesson 8

How do insects' respiratory and circulatory systems function?

Insect Concepts:
- Insects take in oxygen and expel carbon dioxide gas through tiny openings on the sides of their body called spiracles.
- Oxygen taken in by the spiracles moves through a network of tubes called trachea.
- The trachea branch from small to smaller tubes and take oxygen to all cells of an insect's body.
- Trachea also remove used air, carbon dioxide, and take it out through the spiracles.
- Insects are cold-blooded animals, so their body temperatures vary depending on the temperature of the air in their surroundings.
- Insects have an open circulatory system.
- Insect blood is called hemolymph and is clear or greenish in color.

Vocabulary: breathe   closed pumps   oxygen   blood   *trachea (TRAY kee ah)
*spiracles (SPY rah kulz)   *hemolymph (HEE muh limf)

Read: *Lots of Science Library Book #8.*

Insect Systems
Systems
Respiratory system
Circulatory system
Digestive system
Nervous system

Insect Respiratory and Circulatory Systems – Graphic Organizer

Focus Skill: dissecting an insect
Paper Handouts: 3 sheets of 8.5" x 11" paper   a copy of Graphics 8A–D
Graphic Organizer: Make a Hot Dog Layered Look Book with ½" tabs using three sheets of paper. Glue Graphic 8A on the cover of the Layered Look Book and label it *Insect Systems*. This will be referred to as the *Insect Systems Graphic Organizer*. On the first tab, write *Systems*. On the second tab, using a blue marker, write *Respiratory System;* on the third tab, using a red marker, write *Circulatory System*; on the fourth tab, using a purple marker, write *Digestive System;* on the fifth tab, using a green marker, write *Nervous System*.

Open the Layered Look Book to the Systems tab. Cut Graphic 8B into sections. Glue the picture on bottom section. On the top section, glue or write the names of the internal systems of an insect: *Respiratory System, Circulatory System, Digestive System,* and *Nervous System*.

Turn to the Respiratory System tab. Glue Graphic 8C on the bottom section.
✎ Trace the system with a blue marker.
✎✎ Complete ✎. On the top section of the Respiratory System tab, write clue words about the system: *takes in oxygen, releases carbon dioxide, breathes through spiracles and trachea*.
✎✎✎ Complete ✎. On the top section of the Respiratory System tab, explain the system by

using your vocabulary words.
Turn to the Circulatory System tab. Glue Graphic 8D on the bottom section.
- ✎ Trace the system with a red marker.
- ✎✎ Complete ✎. On the top section of the Circulatory System tab, write clue words about the system: *blood is clear or greenish, flows freely in body, heart moves blood.*
- ✎✎✎ Complete ✎✎. On the top section of the Circulatory System tab, explain the system by using your vocabulary words.

Store this Graphic Organizer for use in Lesson Nine.

Insect's Heart – Investigative Loop Lab 8–1

Focus Skill: applying information
Lab Materials: an empty dishwashing liquid container with lid   dishpan or sink deep enough to cover the dishwashing liquid container standing upright
Paper Handouts: a copy of Lab Graphic 8–1   Lab Book   Lab Record Card
Graphic Organizer: Glue Lab Graphic 8–1 on the right pocket of the Lab Book.
Question: How does an insect's heart function?
Research: Read the *Lots of Science Library Book #8* and review the Question.
Prediction: Predict how blood moves inside an insect's body.
Procedure: Fill the dishpan with water. Place the container upright in the water. Squeeze it as if a heart were beating.
Observations: What happened to the water around the container?
Record the Data: On a Lab Record Card, sketch the lab, and draw arrows showing how the water moved around the dishwashing liquid container.
Conclusions: Explain how an insect's heart moves blood in its body. **Possible answer: The insect's heart moves its blood in a similar manner, swishing and swirling the blood around its body.**
Communicate the Conclusions: On a Lab Record Card, compare your observations and conclusions with your predictions. Share your Lab Record Cards with one person who did not participate in this activity.
Spark Questions: Discuss questions sparked by this activity.
New Loop: Choose one question and investigate it further.
- ✎✎✎ **Design Your Own Experiment:** Select a topic based upon the experiences in the *Investigative Loop*. See page viii for more details.

Experiences, Investigations, and Research

Select one or more of the following activities for individual or group enrichment projects. Allow your students to determine the format in which they would like to report, share, or graphically present what they have discovered. This should be a creative investigation that utilizes your students' strengths.

1. Compare and contrast an insect's circulatory system to the human circulatory system.
2. Compare and contrast an insect's respiratory system to the human respiratory system.
3. Most insects' hemolymph does not contain hemoglobin. Some aquatic midges are exceptions. Research this topic.
4. Myrmecology.org (The Scientific Study of Ants)
5. http://www.discovery.com/area/skinnyon/skinnyon970718/skinny1.html

Great Science Adventures

Lesson 9

How do insects' digestive and nervous systems function?

Insect Concepts:
- The insect's digestive system is a tube consisting of a foregut, midgut, and hindgut.
- Digestion takes place primarily in the midgut.
- Digestive systems are specialized to meet insects' particular diets.
- Insects have well-developed nervous systems consisting of a brain, two parallel nerve cords extending from the head to the abdomen, and two fused ganglia in each body segment.
- The fused ganglia, or mini-brains, act independently of the main brain and control the motion of the segment in which they are located.

Teacher's Note: An alternative assessment suggestion for this lesson is found on pages 82-83. If Graphic Pages are being consumed, first photocopy assessment graphics that are needed.

Vocabulary: food   digestion   tube   nerve   *ganglia (GANG lee ah)

Read: *Lots of Science Library Book #9.*

Insect Digestive and Nervous Systems – Graphic Organizer

Focus Skill: explaining the functions of body systems
Paper Handouts: a copy of Graphics 9A–B   *Insect Systems Graphic Organizer*
   All About Insects Graphic Organizer
Graphic Organizer: Open your *Insect Systems Graphic Organizer* to the Digestive System tab. Glue Graphic 9A on the bottom section.
   ✎ Trace the system with a purple marker.
   ✎✎ Complete ✎. Write clue words about the digestive system: *foregut, midgut, hindgut.*
   ✎✎✎ Explain the system using your vocabulary words.

   Turn to the Nervous System tab. Glue Graphic 9B on the bottom section.
   ✎ Trace the system using a green marker.
   ✎✎ Complete ✎. Write clue words about the nervous system: *nerve cords, side by side, brain is in head.*
   ✎✎✎ Explain the system, using your vocabulary words.

   Open your *All About Insects Graphic Organizer.* Glue your *Insect Systems Graphic Organizer* under the Thorax/Abdomen tabs on the bottom section. This completes the *Insect Systems Graphic Organizer* and finalizes the *All About Insects Graphic Organizer.* Display your *All About Insects Graphic Organizer* and entertain questions.

Paper Insect Collection

Focus Skills: graphing, identifying members of a group
Paper Handouts: a copy of Graphics 9C–D
Graphic Organizer: Cut out Graphic 9C. Cut, color, and glue together. Place in one of the boxes of your *Paper Insect Collection*. Cut out Graphic 9D, read the data, and graph the number of species. Place the Insect Data Card in the box.

Insect and Arachnid Bound Book

Focus Skills: graphing, identifying members of a group
Paper Handouts: a copy of Graphics 9C–D
Graphic Organizer: Cut out Graphic 9C. Cut out a dorsal view and a lateral view of the insect. Color and glue on the next page of the *Insect and Arachnid Bound Book*. Cut out Graphic 9D, read the data, graph the number of species, and glue beneath the picture of the insect.

True Insect and Arachnid Collection

Focus Skills: graphing, identifying members of a group
Paper Handouts: a copy of Graphic 9D

Add an aphid to your collection. Read the Insect Data Card, graph the number of species, and pin below your insect. If you cannot find the specified insect, you may mount any insect of your choice. Make your own Insect Data Card.

Human and Insect Sense of Smell

Activity Materials: ripe banana   lemon   honey   vanilla extract   cotton balls
Activity: Soak a cotton ball with the smell of each substance. Keep these in a covered container. Blindfold a partner, remove one cotton ball, and place it under his or her nose. Can your partner guess which one it is? Repeat this with the other cotton balls. Try variations to this activity. Move the cotton balls farther away. Can your partner still smell them? After a while, can your partner still differentiate the smells? Trade places with your partner.
Discuss the Activity: Compare and contrast a human's ability to detect a scent with an insect's ability to detect a scent, based on the observations of this activity and research.

Experiences, Investigations, and Research

Select one or more of the following activities for individual or group enrichment projects. Allow your students to determine the format in which they would like to report, share, or graphically present what they have discovered. This should be a creative investigation that utilizes your students' strengths.

 1. Research the role microorganisms play in an insect's digestive system.

 2. Compare and contrast an insect's nervous system to the human nervous system.

 3. Compare and contrast an insect's digestive system to the human digestive system.

 4. http://yucky.kids.discovery.com/flash/roaches/ (Discovery.com – Yucky Roach World)

Notes

Insects Concept Map
Lessons 10-11
Numbers Refer to Lesson Numbers

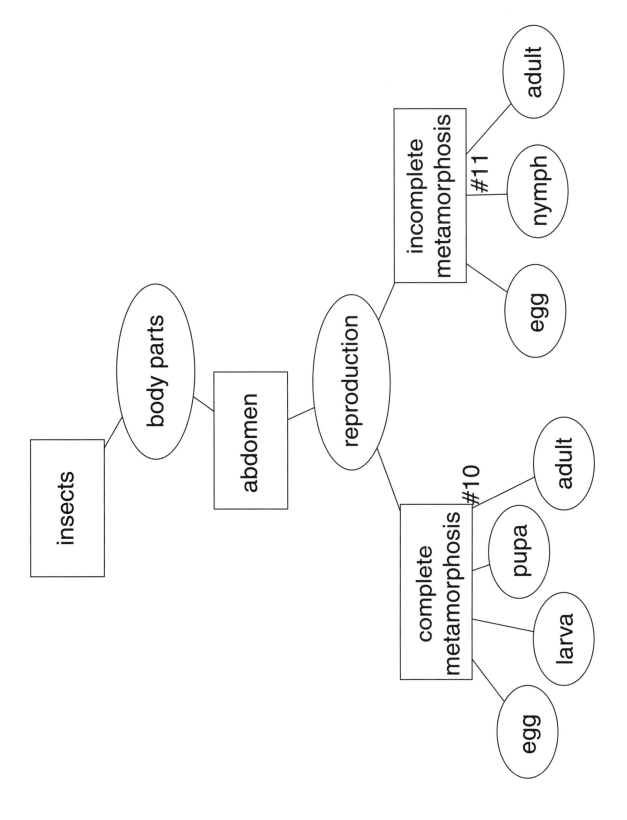

Great Science Adventures

Lesson 10

What is complete metamorphosis?

Insect Concepts:
- Nearly all insects pass through changes in their body form and structure as they grow.
- The process of developing in stages is called metmorphosis.
- There are two types of metamorphosis – complete and incomplete.
- Complete metamorphosis has four stages of development: egg, larva, pupa, and adult.
- Complete metamorphosis always begins with an egg.
- The larva that hatches from the egg looks different from the adult that laid the egg.
- The larval stage is an active period when the young consume great quantities of food.
- After a period of time, the larva enters an inactive period called the pupal stage.
- During the pupal stage, the larva develops into the adult form.
- An adult insect, or imago, emerges from the pupa.

Vocabulary: metamorphosis (met ah MORE fuh sis)   egg   larva   pupa   adult
*imago (i MAY go)   *pupate

Read: *Lots of Science Library Book #10.*

Complete Metamorphosis – Graphic Organizer – Option 1 is a 3D Activity

Focus Skill: communicating information
Paper Handouts: 4 sheets of 8.5" x 11" paper   a copy of Graphics 10A–D   index cards
Graphic Organizer: Make four Pyramid Projects. Cut out Graphics 10A–D and color. Glue each Graphic onto each Pyramid.

✎ Write one name on each of four index cards: *egg, larva, pupa, adult.* Match the card with the correct Pyramid Project.

✎✎ Complete ✎. Choose one insect that goes through complete metamorphosis and describe each stage on a separate index card.

✎✎✎ Complete ✎✎. Research insects and include the names of insects that undergo complete metamorphosis. Choose one insect and narrate its life in the first person.

Glue the Pyramid Projects together to make a Diorama. Review the four stages of complete metamorphosis.

Complete Metamorphosis – Graphic Organizer – Option 2 is a Lay-Flat Activity

Paper Handouts: 8.5" x 11" sheet of paper   a copy of Graphics 10A–D
Graphic Organizer: Make a 4 Door Book. Cut out, color, and glue Graphics 10A–D on each tab of the 4 Door Book. Trim the graphic as needed. Refold it into a Hamburger. Write/copy *Complete Metamorphosis* on the cover. Open the tabs.

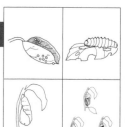

- ✎ Write the name of each stage accordingly: *egg, larva, pupa, adult*.
- ✎✎ Complete ✎. Choose one insect that goes through complete metamorphosis and describe each stage.
- ✎✎✎ Complete ✎✎. Research insects and include the names of insects that undergo complete metamorphosis. Choose one insect and narrate its life in the first person.

Butterfly in Waiting

Activity Materials: cardboard box   plastic wrap   tape   jar   knife

Activity: Find a chrysalis, a pupa case of a butterfly. Cut off a piece of the stem to which the chrysalis is attached. Cut off the lid of a cardboard box. Place the box on its side. Cut a hinged door on the top of the box. Make breathing holes or slits on the sides of the box. Tape plastic wrap over the front opening of the box. Open the hinged top and place the stem and chrysalis inside. Close the hinged top and watch the chrysalis. When the butterfly emerges from the chrysalis, wait until its wings are fully unfurled and dried (about 1–5 hours). Then release the butterfly in the location you found the chrysalis.

Raise Mealworms

Activity Materials: 2 dozen mealworms   flat, plastic container with lid   diced apple
piece of burlap or gauze   window screening   rolled oats
tape

Note: Mealworms are not worms, but the larvae of beetles.

Activity: Put a layer of oats in the bottom of the container. Place the apples on top of the oats. Place half of the mealworms on the apples. Cover this with the burlap. Repeat with layers of oats, apples, mealworms, and burlap. Cut a few holes in the container lid. Place the window screen over the holes and tape it securely in place. Place the lid on the container, place it where it will not be disturbed. Check it after a few days and periodically for the next few weeks.

Note: The larvae will pupate in a few days and become adult beetles in a few weeks in a few weeks. The beetles will then lay eggs, and larvae will appear.

Butterfly and Chrysalis

Paper Handouts: a copy of Graphic 10E
Activity Materials: empty toilet paper tube   black pipe cleaner   popsicle stick
black paint or crayon

Activity: Cut out the butterfly or draw your own butterfly. Make a small hole on the top of the butterfly's head. Insert the pipe cleaner and make a V. Twist it to look like antennae. Glue the popsicle stick on the underside of the butterfly, and let dry. Color the toilet paper tube black to represent the chrysalis. Curl the wings slightly, and insert the butterfly into the chrysalis. Pull out the butterfly with the popsicle stick. Metamorphosis is now complete.

Experiences, Investigations, and Research

Select one or more of the following activities for individual or group enrichment projects. Allow your students to determine the format in which they would like to report, share, or graphically present what they have discovered. This should be a creative investigation that utilizes your students' strengths.

 1. Compare and contrast the "direct development" of animals that are similar to their parents in form and structure at birth, to animals that develop through metamorphosis. Example: Compare and contrast the development of a cow and calf to a butterfly and caterpillar.

 2. Using Graphics 10A–D, make stick puppets. Write a simple play about the metamorphosis of a butterfly and act it out with the puppets.

 3. Research the use of maggots during World War II and in modern–day medicine.

 4. *Who, What, When, Where:* Jan Swammerdam studied insects and defined the different types of metamorphosis.

 5. Research the molting process of an insect.

 6. http://www.ifas.ufl.edu/~pest/vector/chapter_01.htm (University of Florida and the American Mosquito Control Association – Public Health Pest Control)

 7. http://www.discovery.com/area/science/micro/butterfly.html

Notes

Great Science Adventures

Lesson 11

What is incomplete metamorphosis?

Insect Concepts:
- Incomplete metamorphosis has three stages: egg, nymph, and adult.
- A nymph resembles an adult except that it is smaller and has undeveloped wings.
- Nymphs usually eat and live in the same environment as adults.
- Nymphs molt as they grow, gradually looking more and more like an adult.
- Insects vary in the amount of time they live in each stage.

Teacher's Note: An alternative assessment suggestion for this lesson is found on pages 82-83, If Graphic Pages are being consumed, first photocopy assessment graphics that are needed.

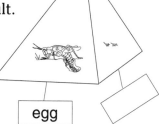

Vocabulary: undeveloped   gills   color   *naiad (NYE ad)   *instar

Read: *Lots of Science Library Book #11.*

Incomplete Metamorphosis – Graphic Organizer   Option 1 is a 3D Activity.

Focus Skill: sequencing a process
Activity Materials: 3 pieces of string, 12" long   hole puncher
Paper Handouts: 8.5" x 11" sheet of paper   3 index cards or 3" x 4" pieces of paper
   a copy of Graphics 11A–C
Graphic Organizer: Make a Pyramid Project. Cut out, color, and glue Graphics 11A–C on each triangle of the Pyramid. On index cards write *egg, nymph, adult*. Match the index card with the correct side of the Pyramid Project. Use string to attach the cards to the Pyramid Project. Attach a string to the top and hang it as a mobile.

Incomplete Metamorphosis – Graphic Organizer   Option 2 is a Lay–flat Activity.

Focus Skill: sequencing a process
Paper Handouts: 8.5" x 11" sheet of paper   a copy of Graphics 11A–C
Graphic Organizer: Make a 3 Tab Book. Cut out, color, and glue Graphics 11A–C on the front of each tab. Open each tab.
✎   Write/copy the words *egg, nymph, adult*.
✎✎   Complete ✎. List insects that go through incomplete metamorphosis.
✎✎✎   Explain each stage of incomplete metamorphosis. Research the process and include new data.

Paper Insect Collection

Focus Skills: graphing, identifying members of a group
Paper Handouts: a copy of Graphics 11 D–E
Graphic Organizer: Cut out Graphic 11D. Cut, color, and glue together. Place in one of the boxes of your *Paper Insect Collection*. Cut out Graphic 11E, read the data, and graph the number of species. Place the Insect Data Card in the box.

Insect and Arachnid Bound Book

Focus Skills: graphing, identifying members of a group
Paper Handouts: a copy of Graphics 11D–E
Graphic Organizer: Cut out Graphic 11D. Cut out a dorsal view and a lateral view of the insect. Color and glue on the next page of the *Insect and Arachnid Bound Book*. Cut out Graphic 11E, read the data, graph the number of species, and glue beneath the picture of the insect.

True Insect and Arachnid Collection

Focus Skills: graphing, identifying members of a group
Paper Handouts: a copy of Graphic 11E

Add a grasshopper to your collection. Read the Insect Data Card, graph the number of species, and pin below your insect. If you cannot find the specified insect, you may mount any insect of your choice. Make your own Data Card.

"The Two Voices" – Graphic Organizer

Paper Handouts: a copy of Graphic 11F   8.5" x 11" sheet of paper
Graphic Organizer: Read the excerpt by Alfred Lord Tennyson:

> Today I saw the dragon-fly
> Come from the wells where he did lie.
> An inner impulse rent the veil
> Of his old husk; from head to tail
> Came out clear plates of sapphire mail.
> He dried his wings; like gauze they grew;
> Through crofts and pastures wet with dew
> A living flash of light he flew.

Make a Half Book and label it *"The Two Voices" by Alfred Lord Tennyson*. Draw a picture under the title. Open the Half Book.

- ✎ Glue Graphic 11F on the bottom section of the Half Book.
- ✎✎ Glue Graphic 11F on the top section. Copy three lines on the bottom section.
- ✎✎✎ On the bottom section, copy the poem. (If you require more writing space, use both the top and bottom sections.)

Read the poetry excerpt to at least one person. Can you sense Tennyson's admiration for the dragonfly?

Experiences, Investigations, and Research

Select one or more of the following activities for individual or group enrichment projects. Allow your students to determine the format in which they would like to report, share, or graphically present what they have discovered. This should be a creative investigation that utilizes your students' strengths.

  1. Compare and contrast complete metamorphosis and incomplete metamorphosis.

  2. List some advantages and disadvantages of incomplete metamorphosis.

  3. Years ago, people in the Far East valued crickets. The wealthy kept singing crickets by their bedsides in ornate cages. Find a cricket, keep it in a ventilated bug jar, and let it serenade you at night.

  4. http://dragonflywebsite.com/ (The Dragonfly Website)

Insects Concept Map
Lessons 12-17
Numbers Refer to Lesson Numbers

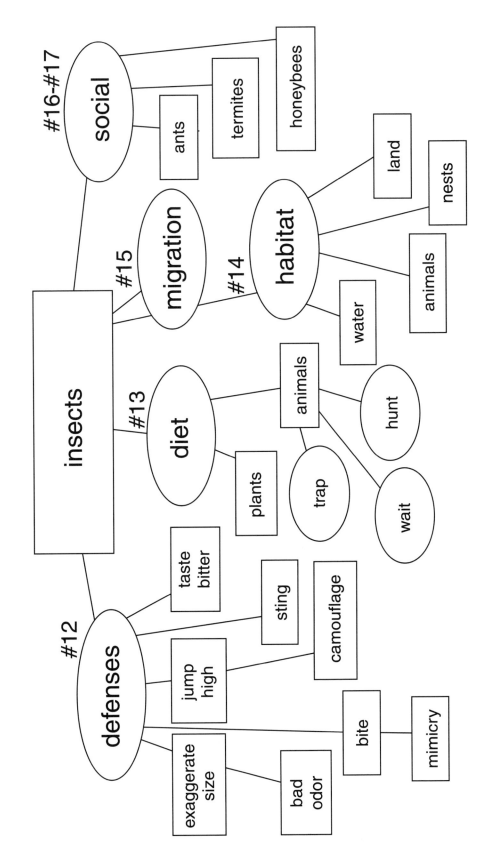

Great Science Adventures

Lesson 12

How do insects defend themselves?

Insect Concepts:
- Insects are a food source for many other animals.
- Insects protect themselves by escaping rapidly, playing dead, assuming threatening body postures, and using defensive weapons.
- Defensive weapons might include the following: pinchers, stingers, poisons, chemical sprays, and unpleasant odors.
- Some insects use camouflage as a means of protection.
- Some insects without protection look like insects that have protective devices. These insects are called mimics.

Vocabulary: food   protect   prey   mimicry   camouflage   *defense   *offense

Read: *Lots of Science Library Book #12.*

Defenses – Graphic Organizer

Focus Skill: cause and effect
Paper Handouts: 2 sheets of 8.5" x 11" paper
   9" x 12" sheet of construction paper
   a copy of Graphics 12A–H
Graphic Organizer: Make two Small Question and Answer Books. Fold the construction paper into a Hot Dog and glue the Small Question and Answer Books inside. On the cover of the construction paper Hot Dog write/copy: *Insect Defenses*. On the tab of the Small Question and Answer Books glue Graphics 12A–H. Under the tabs:
- ✎ Draw a picture of the insect.
- ✎✎ Write clue words describing each means of defense: *scales detach, sounds, exaggerate size, frightening eyes, bite, sting, bad odor, camouflage.*
- ✎✎✎ Identify each insect. Describe the insect's means of defense. Give examples of other insects that use a similar form of defense.

Optional: You may choose other insects to sketch and write about their defenses.

Camouflage

Activity Materials: toothpicks of various colors (red, green, blue, yellow, natural)
Activity: Count 20 toothpicks of each color. Measure off an area of lawn about 4' x 4'. Mark the border with sticks or rocks. Scatter the toothpicks in the designated area. Ask a partner to find as many red toothpicks as possible in 10 seconds. Now, try green toothpicks. Continue with the other colors. Trade places and complete the activity again.

Discuss the Activity: Which toothpicks were the easiest to find? Which were the most difficult to find? Why? What does this tell you about insects and their defenses?

Insect Bat Game

Activity Materials: fabric for a blindfold

Activity: This game is played with at least two people. Mark off an area for play, about 8' x 8'. Choose one person to be the bat. Blindfold the bat. The bat stands in the center of the play area, and the other players, who are insects, stand around him. The bat says "Echo," and the insects must repeat by saying "Echo." The insects may move around the play area, but must respond to the bat's calling. The game ends when the bat has found all the insects.

Leaf Butterfly Drawing

Activity Materials: sheet of paper   crayon

Activity: Find two leaves that are similar in shape. Place the two leaves side by side on the table. Put a little piece of tape on the underside of the leaves to keep them stable. Cover the leaves with the sheet of paper. Hold the paper stable and carefully rub the side of the crayon over the sheet of paper. An image of the leaves will appear. These are the wings of a leaf butterfly. Add the head, eyes, antennae, and body.

Scaly Butterfly Wings

Activity Materials: hole puncher   various colored construction paper   markers   glue

Activity: Fold a yellow piece of construction paper into a Hot Dog. Cut out a shape of a butterfly as shown. Using the hole puncher, punch several colored dots out of construction paper. Open the butterfly and glue the colored dots, which represents scales, on the wings. Design a pattern, overlapping the scales and matching one wing with the other.

Experiences, Investigations, and Research

Select one or more of the following activities for individual or group enrichment projects. Allow your students to determine the format in which they would like to report, share, or graphically present what they have discovered. This should be a creative investigation that utilizes your students' strengths.

1. Research stinging insects. Are the stings acid or base? How are the stings treated?

2. Choose one or more of the insects discussed in this lesson. Research their means of defense. Write a first-person story about the use of the defenses as if you were the insect in danger.

3. http://www.nwf.org/rrick/ (National Wildlife Federation – Ranger Rick)

Great Science Adventures

Lesson 13

What do insects eat?

Insect Concepts:
- Most insects eat plants products such as wood, paper, fabrics, cork, and flour.
- Insects that prey on animals, including other insects, are called predators.
- Predatory insects use inventive methods to catch prey, including trapping, luring, hunting, and ambushing.

Vocabulary: predators lure wait trap *rotate

Read: *Lots of Science Library Book #13.*

Predatory Insects
Insects lure and trap
Insects sit and wait
Insects track and hunt

Predatory Insects – Graphic Organizer

Focus Skill: classifying
Paper Handouts: 2 sheets of 8.5" x 11" paper   a copy of Graphics 13A–D
Graphic Organizer: Make a Layered Look Book. Glue Graphic 13A on the front of the Layered Look Book. Label the cover *Predatory Insects*. Label each tab: *Insects lure and trap, Insects sit and wait,* and *Insects track and hunt.* Glue Graphics 13B–D on the bottom section of each page. On the top section of each page:
- ✎ Draw a picture of each predatory insect.
- ✎✎ Write clue words explaining how predatory insects capture their prey.
- ✎✎✎ Explain how predatory insects capture their prey. Research other predatory insects and include information about those that use similar techniques.

Paper Insect Collection

Focus Skills: graphing, identifying members of a group
Paper Handouts: a copy of Graphics 13E–F
Graphic Organizer: Cut out Graphic 13E. Cut, color, and glue together. Place in one of the boxes of your *Paper Insect Collection*. Cut out Graphic 13F, read the data, and graph the number of species. Place the Insect Data Card in the box.

Insect and Arachnid Bound Book

Focus Skills: graphing, identifying members of a group
Paper Handouts: a copy of Graphics 13E–F

Cut out Graphic 13E. Cut out a dorsal and lateral view of the insect. Color and glue on the next page of the *Insect and Arachnid Bound Book*. Cut out Graphic 13F, read the data, graph the number of species, and glue beneath the picture of the insect.

True Insect and Arachnid Collection

Focus Skills: graphing, identifying members of a group
Paper Handouts: a copy of Graphic 13F

Add a praying mantis to your collection. Read the Insect Data Card, graph the number of species, and pin below your insect. If you cannot find the specified insect, you may mount any insect of your choice. Make your own Data Card.

Praying Mantis	
Class:	Insect
Order:	Mantodea
Number of Species:	1,800
Characteristics:	up to 6 inches long, turns head up to 180 degrees, strong front legs
Habitat:	tropical and subtropical regions
Diet:	insects

Colorful Predators

Activity Materials: hole puncher   colored construction paper: brown, green, red, purple, yellow, orange

Activity: Punch at least 30 holes of each color and mix. Spread the colored dots on a sheet of green paper. You are the predator, the dots are the prey, and the green paper represents the habitat. Have a partner time you to see how many of one color dots you can pick up. You must only pick one dot at a time. Record the color and number of dots. Do this again with other colors and record the number found. Finally, repeat by picking up only green dots. Graph your results. Do you see a marked difference in any of the colors? How do you explain these differences?

Camouflage Patterns

Activity Materials: newspaper   black construction paper   black poster board

Activity: Cut 20 moth patterns out of newspaper and 20 out of black construction paper. Spread newspaper on the floor, and spread the 40 moths on top. Your partner must pick up moths, one at a time, in a 20–second period. He must pick up one moth and stand erect after each one. Count and record. Do the same using black poster board as the background. Trade places with your partner. What did you conclude?

Experiences, Investigations, and Research

Select one or more of the following activities for individual or group enrichment projects. Allow your students to determine the format in which they would like to report, share, or graphically present what they have discovered. This should be a creative investigation that utilizes your students' strengths.

1. Choose a predatory insect to research. Write a newspaper report about the insect and its predatory behavior.

2. Capture an earwig, ladybug, or praying mantis. Look for a plant with aphids. Aphids are often found on new shoots and buds; they are often prolific on rose bushes. Cut off the stem or branch and lay it on a sheet of paper. Place the captured insect beside the branch of aphids. Wait and observe with a magnifying glass.

3. http://www.discovery.com/exp/spiders/spiders.html

Great Science Adventures

Lesson 14

Where do insects live?

Insect Concepts:
- Insects live nearly everywhere.
- They are most abundant in warm climate areas, such as tropical rain forests.
- They are least abundant in arctic regions and ocean habitats.
- The environment in which insects live is called their habitat.
- A specific, smaller part of a habitat is called microhabitat.
- Some insects build nests.

Vocabulary: habitat   nest   water   food   shelter
*microhabitat   *environment

Read: *Lots of Science Library Book #14.*

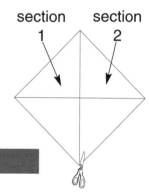

Habitats – Graphic Organizer	Option 1 is a 3D Activity

Focus Skill: classifying
Paper Handouts: 4 sheets of 8.5" x 11 paper   a copy of Graphics 14A–H
Graphic Organizer: Fold four Pyramid Projects, but do not glue them together yet. Cut out and color Graphics 14A–D. On sections 1 and 2 of each Pyramid, glue the habitat Graphics. Now, glue the Pyramid Projects together. Glue the correct Insect Graphic 14E–H in each habitat. Glue all four Pyramid Projects together to make a diorama.

✎✎ On index cards, write the names of other insects that may be found living in that habitat.

✎✎✎ Research each habitat and find other insects that live there. Add them to the correct Pyramid Project. On index cards, record data about each insect and place it in front of the Pyramid Project.

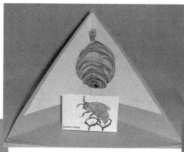

Habitats – Graphic Organizer   Option 2 is a Lay–flat Activity

Focus Skill: classifying
Paper Handouts: 12" x 18" sheet of paper   a copy of Graphics 14A–H
Graphic Organizer: Make a 4 Door Book. Refold the 4 Door Book into a Hamburger. On the front, label it *Habitats*. On the front of each tab, glue parts of Graphics 14A–D. Under the tabs, in the middle section, glue the graphic of the insect that lives in that habitat.
✏️✏️ On the inside of the tabs, write the names of other insects that may be found living in that habitat.
✏️✏️✏️ Research each habitat and find other insects that live there. Add them to the correct tab. Under each tab, record data about each insect.

Wanted Poster

Activity Materials: 8.5" x 11" sheet of paper   pencils   markers
Activity: Make a "Wanted" poster of an insect of your choice. Draw a picture and describe it. The more details you give, the better.

Experiences, Investigations, and Research

Select one or more of the following activities for individual or group enrichment projects. Allow your students to determine the format in which they would like to report, share, or graphically present what they have discovered. This should be a creative investigation that utilizes your students' strengths.

 1. Find a rotten log and carefully roll it over by rolling it towards you. Observe the insects in the microhabitat under the log. Why do you think they live there?

2. Make an insect habitat. Find a jar or aquarium tank and put soil on the bottom. Add sand, rocks, leaves, and twigs. Capture crawling insects and place them gently in their temporary new home. Cover the container with wire mesh and secure it with tape. Observe the insects with a magnifying glass. Release them in a few days.

3. http://www.pbs.org/wgbh/nova/odyssey/hotscience.html (NOVA)

Great Science Adventures

Lesson 15

Why do some insects migrate?

Insect Concepts:
- Some insects hide in small places during cold weather.
- Their blood contains an "antifreeze" that protects them during severe cold.
- Other insects, such as butterflies, migrate during winter.
- Some insects emigrate; they travel to far places in search of food and do not make a return trip.

Teacher's Note: An alternative assessment suggestion for this lesson is found on pages 82-83. If Graphic Pages are being consumed, first photocopy assessment graphics that are needed.

Vocabulary: weather   cold   migration   *overwinter   *emigration

Read: *Lots of Science Library Book #15.*

Paper Insect Collection

Focus Skills: graphing, identifying members of a group
Paper Handouts: a copy of Graphics 15A–C
Graphic Organizer: Cut out Graphic 15A–B. Cut, color, and glue together. Place in one of the boxes of your *Paper Insect Collection*. Cut out Graphic 15C, read the data, and graph the number of species. Place the Insect Data Card in the box.

Insect and Arachnid Bound Book

Focus Skills: graphing, identifying members of a group
Paper Handouts: a copy of Graphics 15A–C
Cut out Graphic 15A. Cut out a dorsal and lateral view of the insect. Color and glue on the next page of the *Insect and Arachnid Bound Book*. Cut out Graphic 15C, read the data, graph the number of species, and glue beneath the picture of the insect.

True Insect and Arachnid Collection

Focus Skills: graphing, identifying members of a group
Paper Handouts: a copy of Graphic 15C
Add a monarch butterfly to your collection. Read the Insect Data Card, graph the number of species, and pin below your insect. If you cannot find the specified insect, you may mount any insect of your choice. Make your own Insect Data Card.

Raise a Monarch Butterfly

Note: This activity is designed for late summer or early fall.
Activity Materials: large glass jar with lid
Activity: In late August or early September, look for milkweed plants growing on the side of the road or in fields. Look for a colorfully striped monarch caterpillar on the leaves. Place milkweed leaves in the jar. Punch holes in the lid and place the caterpillar in the jar. In 3–5 days, the caterpillar will make its way up to the lid and form a chrysalis. After another 10–14 days, the monarch butterfly will emerge from the chrysalis. After the wings have unfurled and hardened, place it on a plant. Watch it fly away.

Mapping Monarch Migration – Graphic Organizer

Focus Skill: mapping
Paper Handouts: 8.5" x 11" sheet of paper   a copy of Graphics 15D–F
Graphic Organizer: Make a Large Question and Answer Book. On the left tab, glue Graphic 15D and using a blue marker, label it *Monarchs West of the Rockies.* On the right tab, glue Graphic 15E and using a red marker, label it *Monarchs East of the Rockies.* Open the tabs. Glue Graphics 15F across the bottom section. Using a blue marker, write on the top section of the left tab, *Where do monarch butterflies west of the Rockies migrate?* Trace the migration route with blue marker. Using a red marker, write on the top section of the right tab, *Where do monarch butterflies east of the Rockies migrate?* Trace the migration route with red marker.

✎✎✎ On the top section of the tabs, describe monarch butterflies' migration patterns. Use your vocabulary words.

Pine Cone Insects

Find an area with pine trees and collect a few pine cones. Over a white sheet of paper, tap the pine cone sharply on the table to remove any insects. Or you may carefully take the pine cone apart. (Careful: some pine cones have sharp edges.) Describe the insects you find. Is a pine cone a good habitat in cold weather? Why or why not?

Experiences, Investigations, and Research

Select one or more of the following activities for individual or group enrichment projects. Allow your students to determine the format in which they would like to report, share, or graphically present what they have discovered. This should be a creative investigation that utilizes your students' strengths.

1. Research monarch butterflies' wintering sites in California and Mexico.

2. Research the migratory habits of Painted Lady butterflies or Bogong moths.

3. http://www.learner.org/jnorth/ – Monarch butterfly migration http://monarchwatch.org/ (University of Kansas Entomology Program – MonarchWatch)

4. http://www.pbs.org/wnet/nature/alienempire/hardware.html (aerodynamics of insects)

5. http://www.insecta.com/ (Spencer Entomological Museum)

Great Science Adventures

Lesson 16

Why are ants and termites called social insects?

Insect Concepts:
- Social insects live and work together for the good of the colony.
- Ants within a colony have specialized roles.
- Workers are wingless, sterile, female ants.
- Workers care for the eggs and developing young, and they maintain the tunnels.
- A winged male ant's purpose is solely to mate with the queen.
- The queen lays all the eggs and she is cared for by the workers.
- Queen ants are born with wings.
- Termite colonies have reproductives, soldiers, and workers.
- Mature male and female reproductives may leave the nest in swarms, mate, and begin new colonies.
- Termite nests are called termitaries. There are two types of termites: those above and below ground.
- Subterranean termites can cause structural damage to buildings.

Vocabulary: colony   social insects   ants   termites   mature   reproduce   *reproductives   *soldiers   *workers   *sterile

Read: *Lots of Science Library Book #16.*

Queen Ant
Worker Ant
Male Ant

Types of Ants – Graphic Organizer

Focus Skills: explaining functions, classifying
Paper Handouts: a copy of Graphics 16A–D   2 sheets of 8.5" x 11" paper
Graphic Organizer: Make a Layered Look Book. Label each tab: *Queen Ant, Worker Ant, Male Ant.* Glue Graphic 16A on the cover and Graphic 16B–D on the bottom of each appropriate page. On the tab section:
- ✏ Draw a picture of each type of ant.
- ✏✏ Write clue words explaining the job of each type of ant: *queen ants lay eggs, worker ants build nest and take care of young, male ants mate with queen*
- ✏✏✏ Describe each ant and explain its functions in the colony, using your vocabulary words.

Ant Farm

Activity Materials: large glass jar   empty soda can   dirt   ants   tape   nylon stocking   rubber band   black construction paper   sand   piece of sponge

Activity: Fill the soda can with sand and tape it shut. Place the can in the glass jar and fill the jar with dirt and ants. The soda can will make the ants build their tunnels against the jar wall so you can view the activity. Dampen the piece of sponge and place it on top of the can.

Place small pieces of food, such as dry pet food, pieces of fruit, or sugar water on the dirt. Cover the jar with the nylon stocking and secure it with a rubber band. Tape black paper over the sides of the jar. Remove the cover only to keep the sponge moist and to replenish food. In about a week, remove the black paper. What do you see? Continue checking each week for several weeks. Return the ants to the location where they were found.

Tracking Termites

Activity Materials: termites   4 sheets of 8.5" x 11" paper   pencil   marker   crayon   Papermate™ blue ink pen

Activity: On separate sheets of paper, draw a large figure eight with each writing tool. Using one sheet at a time, drop a few termites on the sheet. What did you observe? You may need to repeat this several times.

Note: Termites give off pheromones, a special scent. The scent is similar to the ink in a Papermate™ pen, so the termite follows the scent, making a trail.

Poetry – "The Termite" – Graphic Organizer

Paper Handouts: 8.5" x 11" sheet of paper   a copy of Graphic 16E

Graphic Organizer: Read the following poem by Ogden Nash. Make a Half Book. On the cover label it *"The Termite" by Ogden Nash* and draw a picture. Open the Half Book.

> Some primal termite knocked on wood
> And tasted it, and found it good,
> And that is why your Cousin May
> Fell through the parlor floor today.

- On the bottom section, glue Graphic 16E.
- On the bottom section, copy the poem from Graphic 16E.
- Complete . Read the poem to at least one person.

Experiences, Investigations, and Research

Select one or more of the following activities for individual or group enrichment projects. Allow your students to determine the format in which they would like to report, share, or graphically present what they have discovered. This should be a creative investigation that utilizes your students' strengths.

 1. Compare your community to an ant's colony. Think of the different functions and jobs in each. How are they alike? How do they differ?

 2. Draw a diagram of the inside of a termitary.

 3. Find an anthill and place some seeds, grass, alfalfa, or sesame on top. After a few minutes, place small bits of meat. Then scatter sugar grains on the anthill. What did you observe after each type of food was placed on the anthill?

 4. antcam.com (Ant Species Image Archive)

 5. http://www.pbs.org/wgbh/nova/transcripts/2203crea.html (NOVA)

 6. http://www.discovery.com/stories/nature/ants/ants.html

Great Science Adventures

Lesson 17

Why are honeybees called social insects?

Insect Concepts:
- Honeybees are social insects that build and live in hives.
- A honey bee colony includes a queen, workers, and drones.
- Drones are male honeybees whose sole purpose is to mate with the queen.
- Workers are sterile, female honeybees that do all the work in the hive.
- Instead of reproductive parts, workers have stingers and venom to protect the queen and the colony.
- A queen honey bee mates once in her lifetime, storing sperm to use as needed.
- Fertilized eggs produce workers; unfertilized eggs produce drones.

Teacher's Note: An alternative assessment suggestion for this lesson is found on pages 82-83. If Graphic Pages are being consumed, first photocopy assessment graphics that are needed.

Vocabulary: honeybee   hive   honey   stingers   venom   *swarm   *pollen   *drones

Read: *Lots of Science Library Book #17.*

Honeybees
Queen
Drones
Workers

Types of honeybees – Graphic Organizer

Focus Skill: explaining functions
Paper Handouts: 2 sheets of 8.5" x 11" paper   a copy of Graphics 17A–D
Graphic Organizer: Make a Layered Look Book. On the cover of the Book, glue Graphic 17A. Label it *Honeybees*. Label the tabs beginning at the top in the following order: *Queen, Drones,* and *Workers*. Turn to the Queen tab and glue Graphic 17B on the bottom section. On the top section:

- ✎ Draw a picture of the queen bee.
- ✎✎ Write clue words explaining the functions of each bee: *queen lays eggs, drones mate with queen, workers build and protect nest.*
- ✎✎✎ Describe the queen bee and explain its functions in the colony. Use your vocabulary words.

Repeat this procedure for drones and workers, using the next two pages of the Layered Look Book.

Paper Insect Collection

Focus Skills: graphing, identifying members of a group
Paper Handouts: a copy of Graphics 17E–F
Graphic Organizer: Cut out Graphic 17E. Cut, color, and glue together. Place in one of the boxes of your *Paper Insect Collection*. Cut out Graphic 17F, read the data, and graph the number of species. Place the Insect Data Card in the box.

Insect and Arachnid Bound Book

Focus Skills: graphing, identifying members of a group
Paper Handouts: a copy of Graphics 17E–F
Cut out Graphic 17E. Cut out a dorsal and lateral view of the insect. Color and glue on the next page of the *Insect and Arachnid Bound Book*. Cut out Graphic 17F, read the data, graph the number of species, and glue beneath the picture of the insect.

True Insect and Arachnid Collection

Focus Skills: graphing, identifying members of a group
Paper Handouts: a copy of Graphic 17F
Add a honeybee to your collection. Read the Insect Data Card, graph the number of species, and pin below your insect. If you cannot find the specified insect, you may mount any insect of your choice. Make your own Insect Data Card.

Honeybee Syrup

Activity Materials: about 50 clover flowers   2 lb. sugar   1 c. water
1/2 tsp. alum   saucepan   cheesecloth   glass jar with lid

Activity: Collect clover from a field free of pesticides. Place it in a bowl, rinse, and drain. Pour water in the saucepan, add sugar and alum, and boil for about 5 minutes. Pour the hot liquid over the flowers in the bowl. Set aside for about 30 minutes. Strain the liquid through cheesecloth and pour into the glass jar. Your honeybee syrup is ready to use.

Experiences, Investigations, and Research

Select one or more of the following activities for individual or group enrichment projects. Allow your students to determine the format in which they would like to report, share, or graphically present what they have discovered. This should be a creative investigation that utilizes your students' strengths.

 1. *Who, What, When, Where:* Karl Von Frisch, a zoologist, won the Nobel Prize for his study of insects. Research his study on honeybees or another aspect of his studies.

 2. Research royal jelly, the special substance produced by worker bees, which is fed to adult queens.

 3. Research honeybee products such as honey, wax, pollen propolis, royal jelly, and venom, and their uses. Make a graph to show the data found on the production of these products and their uses.

 4. Research the African honeybee, more commonly known as the "killer bee."

 6. http://www.pbs.org/wgbh/nova/bees/ (NOVA)

Notes

Insects Concept Map
Lessons 18-19
Numbers Refer to Lesson Numbers

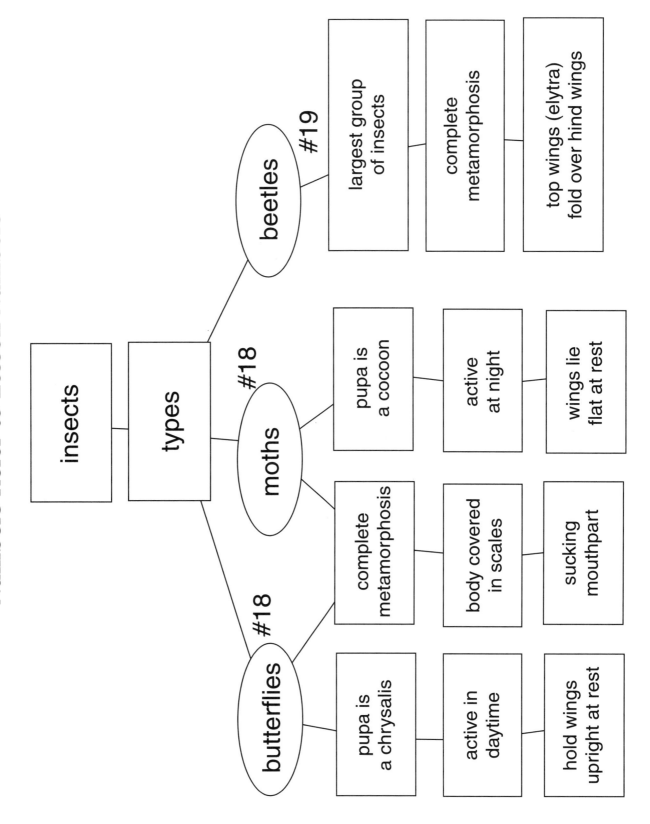

Great Science Adventures

Lesson 18

What are the differences between butterflies and moths?

Insect Concepts:
- Butterflies and moths develop through complete metamorphosis.
- The bodies of butterflies and moths are covered with scales.
- The mouths of butterflies and moths consist of a proboscis, a sucking mouthpart.
- In the larval stage, a caterpillar's job is to eat.
- A butterfly pupa is called a chrysalis.
- Moth caterpillars wrap themselves in a cocoon when they pupate.
- During its larval stage, a Bombyx mori moth is called a silkworm.
- Silkworms secrete a fluid from their spinnerets. The fluid hardens, forming a silky thread.

Vocabulary: butterflies moths caterpillar chrysalis cocoon scales *spinneret *proboscis (proh BOS is)

Read: *Lots of Science Library Book #18.*

Butterfly or Moth – Graphic Organizer

Focus Skill: comparing and contrasting
Paper Handouts: 8.5" x 11" sheet of paper   a copy of Graphics 18A–B
Graphic Organizer: Make a Large Question and Answer Book. Title the cover *Butterflies and Moths*. Color and glue Graphic 18A on the left tab and label it *Butterfly*. Color and glue Graphic 18B on the right tab and label it *Moth*. Open the Butterfly tab. On the bottom section:
- Draw a butterfly.
- Write clue words describing butterflies.
- Describe the characteristics of butterflies. Use your vocabulary words.

Repeat the procedure for the Moth tab.

Catching Butterflies

Activity Materials: butterfly net
Activity: Catching a butterfly requires only a butterfly net and a large amount of patience. Observe a butterfly and sweep the net as shown. When the butterfly is caught in the net, quickly flick your wrist.

Moth Feeding

Activity Materials: 1 lb. brown sugar   2 ripe bananas   beer   wide–tip paintbrush   flashlight
Activity: In a bowl, mash the bananas and add the sugar and beer. Mix well. Apply the mixture with a paintbrush to the bark of a few trees, in an area about 3" x 12". Later in the evening, look at the trees with your flashlight.

Butterfly Garden

Activity Materials: butterfly–attracting plants such as ageratum, beebalm, bougainvillaea, calendula, coneflower, dahlia, daylily, geranium, hibiscus, impatiens, marigold, milkweed, mint family, phlox, salvia, snapdragon, yarrow, yellow sage, zinnia
Activity: The best spot for a butterfly garden is a sunny place, away from wind, and against tall bushes or a wall. Ask an informed employee at a nursery or garden shop for tips on planting the flowers. Avoid pesticides.

Cocoons

Activity: Go outdoors and look for cocoons. They are plentiful and can be found year round. Look on tree branches, leaves, bushes, windowsills, and sheds. If the cocoon is attached to a branch, carefully cut off the branch. Place it in the jar. Cover the jar with the screen and secure it with a rubber band. Place the jar in a safe place that is fairly consistent with the outdoor temperature, such as in the garage or on a porch. A moth should emerge in the spring or summer. Release the moth a few hours after its wings have fully dried.

Experiences, Investigations, and Research

Select one or more of the following activities for individual or group enrichment projects. Allow your students to determine the format in which they would like to report, share, or graphically present what they have discovered. This should be a creative investigation that utilizes your students' strengths.

 1. Visit a butterfly farm.

 2. Look for butterflies and moths. Observe the position of their wings at rest. Were they flying during the day or at night? Describe their antennae. Describe their color and markings.

 3. http://butterflywebsite.com/ (The Butterfly Website)

Great Science Adventures

Lesson 19

What are beetles?

Insect Concepts:
- One out of every five animal species is a beetle.
- Their numbers prove that beetles are one of the most prolific animals to ever live.
- Beetles live everywhere and feed on all kinds of plants, and they eat both living and dead animals.
- Beetles undergo complete metamorphosis.
- Some beetles are beneficial; others are considered to be pests.
- Beetles are identified by their top wings, called elytra, which their hindwings fold underneath.

Vocabulary: beetles   successful   *coleoptera (ko lee OP ter ah)   *elytra (ee LIE trah)

Read: *Lots of Science Library Book #19.*

Paper Insect Collection

Focus Skills: graphing, identifying members of a group
Paper Handouts: a copy of Graphics 19A–B
Graphic Organizer: Cut out Graphic 19A. Cut, color, and glue together. Place in one of the boxes of your *Paper Insect Collection*. Cut out Graphic 19B, read the data, and graph the number of species. Place the Insect Data Card in the box. This completes your *Paper Insect Collection*. Display your *Paper Insect Collection* and make a simple oral presentation.

Insect and Arachnid Bound Book

Focus Skills: graphing, identifying members of a group
Paper Handouts: a copy of Graphics 19A–B
  Cut out Graphic 19A. Cut out a dorsal and lateral view of the insect. Color and glue on the next page of the *Insect and Arachnid Bound Book*. Cut out Graphic 19B, read the data, graph the number of species, and glue beneath the picture of the insect. Make a simple oral presentation using your Bound Book.

True Insect and Arachnid Collection

Focus Skills: graphing, identifying members of a group
Paper Handouts: a copy of Graphic 19B

Add a beetle to your collection. Read the Insect Data Card, graph the number of species, and pin the card below your insect. If you cannot find the specified insect, you may mount any insect of your choice. Make your own Data Card and add information. Although you will continue to add arachnids to your *True Insect and Arachnid Collection,* display it and make a simple oral presentation.

Beetles – Graphic Organizer

Focus Skill: describing a process
Paper Handouts: 2 sheets of 8.5" x 11" paper   a copy of Graphics 19C–F
Graphic Organizer: Make a Bound Book. Glue Graphic 19C on the front of the Bound Book and label it *Beetles*. Turn the page and glue Graphic 19D on the righthand side. Repeat for the next two pages with Graphics 19E–F. Turn to the first page:

- ✏ On the lefthand side, draw the stages of a beetle's life. Turn the page. On the lefthand side, draw a beetle necklace. Turn the page. On the lefthand side, draw a beetle with both pairs of wings open in flight.

- ✏✏ On the lefthand side, write clue words about a beetle's metamorphosis: *egg, larva, pupa, adult*. Turn the page, and on the lefthand side, write clue words about the colors of beetles: *black, green, blue, red*. Turn the page, and on the lefthand side, write clue words about the wings of beetles: *top wings-elytra, hard wings that protect hind wings, hind wings folded underneath*.

- ✏✏✏ On the lefthand side, describe the metamorphosis of a beetle. Turn the page and describe how beetles appear and ways that they are used. Turn the page and describe the process of a beetle's flight.

 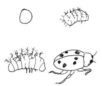

Beetle Trails

Activity Materials: food coloring   paper plate   sheet of paper   tweezer
Activity: Find a beetle, place it in a bug jar, and put it in the fridge for 15 minutes. The cooler temperature will not kill the insect but will slow it down. Place two drops of food coloring on the plate. Release the beetle on the food coloring so its feet get wet. With tweezers, gently place it on the sheet of paper and let the beetle make a trail. What does the trail look like? Try this again with other beetles. Compare the trails.

Ladybug Magnet

Activity Materials: small, smooth rock   black pipe cleaner   poster paint
        magnet strip   glue
Activity: Look at a picture of a ladybug. Paint the ladybug's pattern on the rock. Cut pieces of pipe cleaner and glue on for legs and antennae. When dry, glue the magnetic strip on the back.
Optional: You might want to glue on plastic eyes, which are readily available at craft stores.

Experiences, Investigations, and Research

Select one or more of the following activities for individual or group enrichment projects. Allow your students to determine the format in which they would like to report, share, or graphically present what they have discovered. This should be a creative investigation that utilizes your students' strengths.

1. Go on a scavenger hunt in a field or vacant lot. Try to locate at least one of each of the following:
   Arthropod with six legs
   Arthropod with eight legs
   Arthropod with no wings
   Arthropod with one pair of wings
   Arthropod with two pairs of wings
   Arthropod caught in a web
   Arthropod of more than two colors
   Arthropod during its larval stage
   Arthropod during its pupal stage
   Arthropod that undergoes complete metamorphosis
   Arthropod that undergoes incomplete metamorphosis

2. http://www.pbs.org/wgbh/nova/kalahari/beetle.html (NOVA)

Notes

Great Science Adventures

Lesson 20

Are insects helpful or harmful?

Insect Concepts:
- Ninety–nine percent of all insects are beneficial to human-kind.
- Insects add beauty to our world and are a food source for many animals.
- Insects destroy pests, pollinate crops, and provide honey.
- Insects break down decaying matter and animal droppings.
- Harmful insects include mosquitoes, termites, and aphids.
- Insects can transmit disease and destroy structures and crops.

Vocabulary: harmful   beneficial   predatory   pests   crops   *pesticides

Read: *Lots of Science Library Book #20.*

Helpful Insects and Harmful Insects – Graphic Organizer

Focus Skill: compare and contrast
Paper Handouts: 8.5" x 11" sheet of paper   a copy of Graphics 20A–C
Graphic Organizer: Make a Large Question and Answer Book. Glue Graphics 20A on the cover of the Large Question and Answer Book. Open the book. Label one tab *Helpful Insects* and the other tab *Harmful Insects*. Open the Helpful Insects tab and glue Graphics 20B on the bottom section. On the top section:
- ✏ Draw pictures of helpful insects.
- ✏✏ Write clue words explaining how insects are helpful.
- ✏✏✏ List examples of helpful insects. Explain how they are helpful.

Repeat the procedure for the Harmful Insects tab using Graphics 20C.

Aphids – Investigative Loop   Lab 20-1

Note: This activity is best completed in the spring when aphids are plentiful.
Focus Skill: observing
Lab Materials: jar filled with water   magnifying glass
Paper Handouts: 8.5" x 11" sheet of paper   Lab Book   Lab Record Cards
    a copy of Lab Graphic 20–1
Graphic Organizer: Make a Pocket Book. Glue it side-by-side to the Lab Book. Glue Lab Graphic 20–1 on the left pocket.
Concept: Aphids are insects.
Research: Review *Lots of Science Library Books #3 – #7* and review the characteristics of insects.
Predictions: Since aphids are insects, what qualitative characteristics do you expect to find when observing them? Write your predictions on a Lab Record Card, labeled "Lab 20–1."

Procedure: Look for a plant with aphids. Aphids are often found on new shoots and buds, especially on rosebushes. Cut off a small branch that contains them. Immerse the entire branch in the jar of water. This will cause the aphids to hold tightly to the branch. Remove the branch.

Observations: Observe the aphids with a magnifying glass. Can you locate the head, thorax, abdomen, and antennae? Are all the aphids the same size? Can you see them sucking the plant juices?

Record the Data: Label a Lab Record Card, "Lab 20–1." Draw an aphid you observed. Label the parts of the aphid. List the insect characteristics observed in the lab.

Conclusions: Draw conclusions about aphids, based on this lab.

Communicate the Conclusions: Label a Lab Record Card "Lab 20–1." Write your conclusions about aphids. Defend the conclusions with the data.

Spark Questions: Discuss questions sparked by this lab.

New Loop: Choose one question to investigate further.

✎✎✎ **Design Your Own Experiment:** Select a topic based upon the experiences in the *Investigative Loop*. See page viii for more details.

Lay a Trail

Focus Skill: applying information
Activity Materials: sugar water
Activity: Find a spot outside where ants live. Pour a little sugar water on the ground. Check it every two hours.
Note: You should observe ants discovering the bait. They will hurry back to their nest, dragging their abdomens along the ground. Other ants will discover this invisible trail with their antennae. As more ants travel this trail, the scent becomes stronger. Soon, ants will be running rapidly to and fro.

Poetry – "The Fly" – Graphic Organizer

Paper Handouts: 8.5" x 11" sheet of paper
Graphic Organizer: Read and discuss the following poem by Ogden Nash. Make a Half Book, label the front of it *"The Fly" by Ogden Nash,* and draw a picture. Open the Half Book. On the bottom section, copy the following poem.

> God in his wisdom made the fly
> And then forgot to tell us why.

Experiences, Investigations, and Research

Select one or more of the following activities for individual or group enrichment projects. Allow your students to determine the format in which they would like to report, share, or graphically present what they have discovered. This should be a creative investigation that utilizes your students' strengths.

 1. Research the bubonic plague.

 2. About 50% of insecticide use in the U.S. is within the tobacco and cotton industries. Research one of these crops, its pests, and how the pests are controlled.

 3. Research biological control.

 4. Choose one insect and investigate its importance in our environment.

 5. http://www.orkin.com/Main.htm (Orkinpests)

 6. http://www.discovery.com/area/science/micro/mosquito.html

Insects Concept Map
Lessons 21-24
Numbers Refer to Lesson Numbers

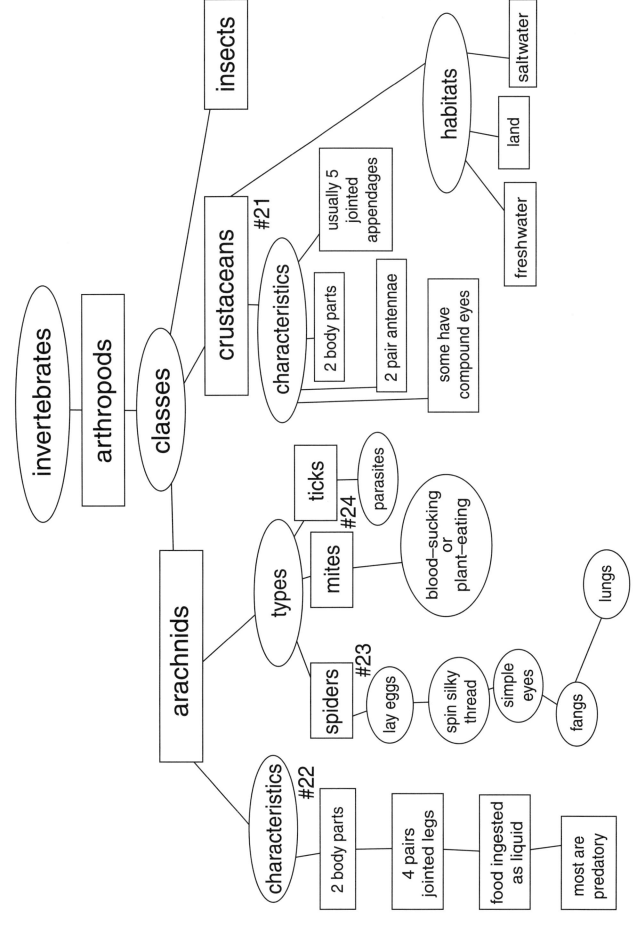

Great Science Adventures

Lesson 21

What are crustaceans?

Anthropod Concepts:
- Arthropods consist of insects, crustaceans, and arachnids.
- Most crustaceans live in freshwater or saltwater; pillbugs live on land.
- Crustaceans have an exoskeleton, two segmented body parts, and two pairs of antennae.
- They usually have five pairs of jointed appendages; one pair may be modified to form pincers.
- Crustaceans such as krill and copepods, are very small and are an important link in the food web.
- Larger crustaceans, such as crabs and lobsters, have compound eyes.

Vocabulary: crustaceans   crab   lobster   krill   *cephalothorax (sef eh le THOR ax) *chelae (KEE lee)

Read: *Lots of Science Library Book #21.*

Lobsters are Crustaceans – Graphic Organizer

Focus Skill: labeling parts
Paper Handouts: 8.5" x 11" sheet of paper   a copy of Graphic 21A
Graphic Organizer: Make a Half Book. Glue Graphic 21A on the cover of the Half Book and title it *Lobsters are Crustaceans.*

✎   Open the Half Book. On the bottom section, draw pictures of other animals that live in the ocean with lobsters.

✎✎   On the front of the Half Book, label the following parts: *compound eyes, stalk, antennae, cephalothorax, abdomen, legs, and pincers.* Open the Half Book. On the bottom section, write clue words explaining crustaceans in general or specifically about lobsters.

✎✎✎   On the front of the Half Book, label the following parts: *compound eyes, facets, stalk, antennae, antennules, cephalothorax, abdomen, legs, and pincers.* Research the lobster and crab industries and write a short paragraph about them on the bottom section. If you require more writing space, you may use both the top and bottom sections.

Experiences, Investigations, and Research

Select one or more of the following activities for individual or group enrichment projects. Allow your students to determine the format in which they would like to report, share, or graphically present what they have discovered. This should be a creative investigation that utilizes your students' strengths.

 1. Research how krill and copepod play a major role in the food chain.

 2. http://www.crustacea.net/ (An Information Retrieval System for Crustaceans of the World)

Great Science Adventures

Lesson 22

What are arachnids?

Arachnid Concepts:
- Arachnids have two segmented body parts and four pairs of jointed legs, covered with an exoskeleton.
- Arachnids do not have chewing or biting mouthparts; food is ingested as a liquid.
- Arachnids do not have antennae or wings.
- Most arachnids are predatory.
- Examples of arachnids include scorpions, spiders, mites, and ticks.
- Some arachnids produce a painful, sometimes dangerous, sting or bite.
- Scorpions are carnivores; most mites are parasites.

Vocabulary: arachnids   liquid   *book lungs   *chelicerae (key LISS er ee)   *pedipalps (ped ee palps)   *carnivores   *parasite

Read: *Lots of Science Library Book #22.*

Arachnids and Insects – Graphic Organizer

Focus Skill: compare and contrast
Paper Handouts: 8.5" x 11" sheet of paper   a copy of Graphics 22A
Graphic Organizer: Make a Half Book. Glue Graphic 22A on the cover. Cut on the dotted lines, through the cover paper only, to make a 3 Tab Book. Label the left side *Insects,* the right side *Arachnids*, and the middle *Insects and Arachnids*. On the bottom section of the left tab, write clue words common to only insects: *six jointed legs, three body parts*. On the bottom section of the right tab, write clue words common only to arachnids: *eight jointed legs, two body parts*. On the bottom section of the middle tab, write clue words that are common to both insects and arachnids: *jointed legs, segmented bodies*.

Arachnid Collection – Option 1 – Graphic Organizer

Focus Skills: graphing, identifying members of a group
Paper Handouts: 3 sheets of construction paper   a copy of Graphics 22B–C
Graphic Organizer: Make three Display Box Organizers, and glue them together as shown. You will be adding to this collection through Lesson 24. This will be referred to as the *Paper Arachnid Collection*. Cut out Graphic 22B, color, fold, and glue. Place the arachnid in one of the boxes in your *Paper Arachnid Collection*. Cut out Graphic 22C, read the data, graph the number of species, and glue the Arachnid Data Card in the appropriate box.

Arachnid Collection – Option 2 – Graphic Organizer

Focus Skills: graphing, identifying members of a group

Paper Handouts: *Insect and Arachnid Bound Book* a copy of Graphics 22B–C

Graphic Organizer: Cut out Graphic 22B. Cut out a dorsal view and a lateral view of the arachnid. Color and glue on the next page of the *Insect and Arachnid Bound Book.* Cut out Graphic 22C, read the data, graph the number of species, and glue the Data Card beneath the picture of the arachnid. This will continue to be referred to as the *Insect and Arachnid Bound Book.*

Arachnid Collection Option 3

Focus Skills: graphing, identifying members of a group
Paper Handouts: a copy of Graphic 22C

Continue using your *True Insect and Arachnid Collection.* Add a scorpion to your collection. Read the Arachnid Data Card, graph the number of species, and pin below your insect. If you cannot find the specified arachnid, you may mount any arachnid of your choice. Make an Arachnid Data Card. This will continue to be referred to as the *True Insect and Arachnid Collection.*

Arachnid Body Parts

Activity Materials: egg carton   4 black pipe cleaners
Activity: Cut two egg carton sections as one piece. Pierce the pipe cleaners through the abdomen to make eight legs. With black permanent marker, draw the eyes and mouth on the cephalothorax. Review the parts of an arachnid.

Tasty Spider

Activity Materials: drinking glass   bread   peanut butter   pretzels   raisins
Activity: Use a drinking glass upside down to cut out a circle of bread. Spread with peanut butter. Roll some of the remaining bread in your palm and form a ball for the cephalothorax. Stick 8 pretzels in the abdomen for legs. Cut the raisins into small bits. Place 8 pieces of raisins on the spider's head for eyes. Eat your spider.

Experiences, Investigations, and Research

Select one or more of the following activities for individual or group enrichment projects. Allow your students to determine the format in which they would like to report, share, or graphically present what they have discovered. This should be a creative investigation that utilizes your students' strengths.

1. Research scorpions. Do scorpions live in your area? Find out information about Arizona bark scorpions.

2. Find out how arachnids are helpful to man.

3. Look for dead insects on spider webs. Spiders often wrap their prey with silk. Using tweezers, carefully remove the dead insect. Unwrap the silk threads carefully. Do you think the spider has already sucked out the insect's fluids? Why?

4. Spiders are carnivores. Define herbivore, omnivore, and insectivore.

5. http://www.ufsia.ac.be/Arachnology/Pages/Kids.html (Download a free spider screen saver)

Great Science Adventures

Lesson 23

What are spiders?

Arachnid Concepts:
- Spiders have two body parts and four pairs of jointed legs, covered with an exoskeleton.
- Spiders lay eggs and their young do not go through metamorphosis.
- All spiders spin a silky thread, but not all spiders build a web.
- Spiders can move long distances by ballooning, or allowing the wind to catch their silky strands.
- Spiders secrete a special oil on their feet to prevent them from sticking on their own web.
- Spiders construct webs in a variety of shapes: orb, triangle, tangle, and sheet webs.
- Spiders have only simple eyes and cannot see well.
- Spiders' mouthparts consist of fangs for injecting poison.
- Spiders breathe with two or four lungs.
- The larger female spider often eats the male after mating.
- Spiders are beneficial because they eat crop-destroying and disease-transmitting insects.

Vocabulary: spider web   oil   nest   *ballooning   *spiderlings

Read: *Lots of Science Library Book #23.*

Paper Arachnid Collection – Graphic Organizer

Focus Skills: graphing, identifying members of a group
Paper Handouts: a copy of Graphics 23A–B
Graphic Organizer: Cut out Graphic 23A, color, glue, and fold. Place the arachnid in one of the boxes. Cut out Graphic 23B, read the data, graph the number of species, and glue the Data Card on the box.

Insect and Arachnid Bound Book – Graphic Organizer

Focus Skills: graphing, identifying members of a group
Paper Handouts: a copy of Graphics 23A–B
Cut out Graphic 23A. Cut out a dorsal view and a lateral view of the arachnid. Color and glue on the next page of the *Insect and Arachnid Bound Book*. Cut out Graphic 23B, read the data, graph the number of species, and glue the Data Card below the picture of the arachnid.

True Insect and Arachnid Collection

Focus Skills: graphing, identifying members of a group
Paper Handouts: a copy of Graphic 23B
Add a spider to your collection. Cut out the Arachnid Data Card and read the information.

Graph the number of species, and pin below the arachnid. If you cannot find the specified arachnid, you may mount any arachnid of your choice. Make an Arachnid Data Card.

Spider – Graphic Organizer

Focus Skill: communicating information
Paper Handouts: 8.5" x 11" sheet of paper   a copy of Graphic 23C
Graphic Organizer: Make a Trifold Book. Glue Graphic 23C on the cover. Open the Trifold Book.
- ✏ Draw pictures of spiders.
- ✏✏ On the top section, complete ✏. On the bottom section, write clue words describing spiders. On the middle section, sketch a spider web.
- ✏✏✏ On the top section, complete ✏. On the middle section, sketch the different types of spider webs and identify them. On the bottom section, describe the characteristics of spiders. Use your vocabulary words.

Spider's Web

Activity Materials: spray bottle containing water or garden hose
Activity: Observe spider webs. Sketch what they look like. Spray them gently with water to make them easier to see.

Ballooning Spider

Activity Materials: sheet of paper   tape   drinking straw
    10 black pipe cleaners   fishing wire or string
Activity: Fold the sheet of paper into a Hot Dog and cut out a spider as shown. Unfold and color the spider. On the other side, cut a piece of straw to run the length of the abdomen. Tape the straw as shown. Cut pipe cleaners in half. Tape the pipe cleaners on the underside of the abdomen to make legs. Thread the fishing wire or string through the straw. Cut a ½–inch piece of straw and tie it to one end of the wire. Tie another ½–inch piece of straw to the other end. Now, hold the spider at the top of the string and release it. Watch the spider balloon.

Unsticky Spider Feet – Investigative Loop

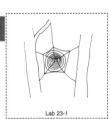
Lab 23-1

Focus Skill: experiencing a concept
Lab Materials: clear tape   cooking oil
Paper Handouts: Lab Book   a copy of Lab Graphics 23–1   Lab Record Cards
Graphic Organizer: Glue Lab Graphic 23–1 on the right pocket
Question: Why don't a spider's feet stick to its own web?
Research: Read the *Lots of Science Library Book #23* and review the Question.
Procedure: Roll a piece of clear tape into a circle. Place it on a table. Pretend your fingers are feet and walk across the tape. Roll another piece of clear tape into a circle. Place it on a table. Dip your fingers in the oil and walk across the tape.
Observations: What happened when your fingers walked over the sticky side of the tape? What happened when your oil–dipped fingers walked over the sticky side of the tape?
Record the Data: Label two Lab Record Cards "Lab 23–1." On one card, describe how the tape felt on your fingers. On another card, describe how the tape felt with oil–dipped fingers.
Conclusions: Review your Lab Record Cards. Draw conclusions about why a spider's feet don't stick to its web.

Communicate the Conclusion: Label a Lab Record Card "Lab 23–1." Write your conclusions on the card.
Spark Questions: List questions that were sparked by this activity.
New Loop: Choose a sparked question to investigate further.
   ✎✎✎ **Design Your Own Experiment:** Select a topic based upon the experiences in the *Investigative Loop*. See page viii for more details.

Experiences, Investigations, and Research

Select one or more of the following activities for individual or group enrichment projects. Allow your students to determine the format in which they would like to report, share, or graphically present what they have discovered. This should be a creative investigation that utilizes your students' strengths.

 1. Find a spider web and spray it with white spray paint. Hold a stiff sheet of black paper with both hands and push the paper into the web. Ask your partner to snip the silk threads at the edges of the paper. Look at the web. You may want to do this with several different kinds of webs and compare them.
Optional: Use black spray paint and white paper.
Note: This activity is best completed in the morning.

 2. Research black widow spiders and brown recluse spiders.

 3. Locate a spider web. Test the strength of a strand of web. Try strands from other webs.

 4. Capture a live spider. Place a pencil in front of it and let it walk onto the pencil. Let, then walk to the end of the pencil, then gently push it off. Observe how the spider moves on the strand of silk.

 5. http://www.reptile-world.com/

Notes

Great Science Adventures

Lesson 24

What are mites and ticks?

Arachnid Concepts:
- Mites make up the largest group of arachnids.
- Most mites are blood-sucking arachnids, but some live off plants.
- Common house dust mites may cause asthma in humans.
- Ticks are external parasitic arachnids; they feed solely on blood.
- There are two kinds of ticks: hard and soft ticks.
- The more common hard tick is mainly found on mammals and is tear–shaped.
- Ticks transmit many harmful diseases.

Teacher's Note: An alternative assessment suggestion for this lesson is found on pages 82-83. If Graphic Pages are being consumed, first photocopy assessment graphics that are needed.

Vocabulary: mite   tick   blood   oval   tear   *scutum (SKEW tum)
*capitulum (kah PITCH u lum)   *host   *questing   *asthma

Read: *Lots of Science Library Book #24.*

Ticks – Graphic Organizer

Focus Skill: describing types of a group
Paper Handouts: 8.5" x 11" sheet of paper   a copy of Graphics 24A
Graphic Organizer: Make a Half Book. Label the cover of the book *Hard Ticks and Soft Ticks*. Glue Graphics 24A on the cover. Inside the book:
- ✎ Draw each type of tick.
- ✎✎ Write clue words describing ticks: *parasites, hard, soft, live on animals or plants.*
- ✎✎✎ Describe the characteristics of ticks. List examples of ticks. On the top section, explain the life cycles of each. Use your vocabulary words.

Paper Arachnid Collection – Graphic Organizer

Focus Skills: graphing, identifying members of a group
Paper Handouts: a copy of Graphics 24B-C
Graphic Organizer: Cut out Graphic 24B, color, glue, and fold. Place the arachnid in one of the boxes. Cut out 24C, read the data, graph the number of species and glue the Data Card on the box. This completes your *Paper Arachnid Collection*. Present your *Paper Arachnid Collection* and entertain questions.

Insect and Arachnid Bound Book – Graphic Organizer

Focus Skills: graphing, identifying members of a group
Paper Handouts: a copy of Graphics 24B–C

Cut out Graphic 24B. Cut out a dorsal and a lateral view of the arachnid. Color and glue on the next page of the *Insect and Arachnid Collection Bound Book*. Cut out Graphic 24C, read the data, graph the number of species, and glue the Data Card below the picture of the arachnid. This completes your *Insect and Arachnid Bound Book*. Present your *Insect and Arachnid Bound Book* and entertain questions.

True Insect and Arachnid Collection

Focus Skills: graphing, identifying members of a group
Paper Handouts: a copy of Graphic 24C

Add a tick to your collection. Cut out the Arachnid Data Card and read the information. Graph the number of species, and pin the cord below the arachnid. If you cannot find the specified arachnid, you may mount any arachnid of your choice. Make an Arachnid Data Card. This completes your *True Insect and Arachnid Collection*. Present your *True Insect and Arachnid Collection* and entertain questions.

Experiences, Investigations, and Research

Select one or more of the following activities for individual or group enrichment projects. Allow your students to determine the format in which they would like to report, share, or graphically present what they have discovered. This should be a creative investigation that utilizes your students' strengths.

 1. Research at least one disease transmitted by ticks.

 2. Discover ways in which animals try to rid themselves of ticks and mites.

 3. http://www.discovery.com/area/science/micro/tick.html

Great Science Adventures

Assessment: An Ongoing Process

Students do not have to memorize every vocabulary word or fact presented in these science lessons. It is more important to teach them general science processes and cause and effect relationships. Factual content is needed for students to understand processes, but it should become familiar to them through repeated exposure, discussion, reading, research, presentations, and a small amount of memorization. You can determine the amount of content your students have retained by asking specific questions that begin with the following words: *name, list, define, label, identify, draw,* and *outline*.

Try to determine how much content your students have retained through discussions. Determine how many general ideas, concepts, and processes your students understand by asking them to describe or explain them. Ask leading questions that require answers based on thought and analysis, not just facts. Use the following words and phrases as you discuss and evaluate: *why, how, describe, explain, determine,* and *predict*. Questions may sound like this:

What would happen if _____?   *Compare _____ to _____.*
Why do you think _____ happens?   *What does ___have in common with __?*
What do you think about _____?   *What is the importance of _____?*

Alternative Assessment Strategies

If you need to know specifically what your students have retained or if you need to assign your students a grade for the content learned in this program, we suggest using one of the following assessment strategies.

By the time your students have completed a lesson in this program, they will have written about, discussed, observed, and discovered the concepts of the lesson. However, it is still important for you to review the concepts that you are assessing prior to the assessment. By making your students aware of what you expect them to know, you provide a structure for their preparations for the assessments.

1) At the end of each lesson, ask your students to restate the concepts taught in the lesson. For example, if they have made a 4 Door Book showing the steps of a process, ask them to tell you about each step, using the pictures as a prompt. This assessment can be done by you or by a student.

2) At the end of each lesson, ask your students to answer the questions on the inside back cover of the *Lots of Science Library Book* for that lesson. The answers to these questions may be given verbally or in writing. Ask older students to use their vocabulary words in context as they answer the questions. This is a far more effective method of determining their knowledge of the vocabulary words than a matching or multiple choice test on the words.

3) Provide your students with Concept Maps that have been partially completed. Ask them to fill in the blanks. Example for after Lesson 7:

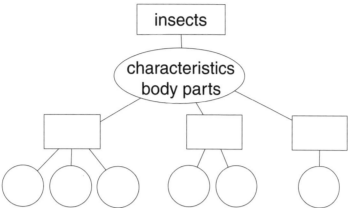

4) Use the 3D Graphic Organizers to assess your students' understanding of the concepts. Use the Concepts listed on the teacher's pages to determine exactly what you want covered in the assessments. Primary and beginning students may use the pictures found on the Graphics Pages as guides for their assessments. By using the pictures, your students are sequencing and matching while recalling information. Older students should draw their own pictures and use their vocabulary words in their descriptions of the concepts. Below are suggestions for this method of assessment.

   a) Lesson 7 - Make a 3 Tab Book. On the cover, sketch a generic insect with each body part on a tab. Include all external parts of the insect. Under each tab, describe the insect and explain the purposes of specific parts.

   b) Lesson 9 - Make a 4 Door Book. On each tab, write the name of an internal system of insects. Under the tab, explain how the system operates.

   c) Lesson 11 - Make a Large Question and Answer Book. On the left tab, write *Complete Metamorphosis* and on the right tab, write *Incomplete Metamorphosis*. Under the tabs, explain the process of each, including the advantages and disadvantages. On the back of the book, compare and contrast the two types of metamorphoses.

   d) Lesson 15 - Make a Half Book. Choose one of the following topics for the book. Title the cover, sketch an illustration, and inside explain or describe: 1) How and why insects defend themselves; 2) How predatory insects survive; 3) Where do insects live? 4) Which insects migrate, and why?

   e) Lesson 17 - Make a 2–paper Layered Look Book. Choose one social insect and explain the purpose of each member of the colony. Describe their duties, purposes, and importance for the survival of the group.

   f) Lesson 24 - Make a 3 Tab Book. Title the tabs *Insects, Crustaceans, and Arachnids*. Under each tab, list characteristics of that group. On the back of the book, compare and contrast the three types of arthropods.

Notes

Great Science Adventures

Lots of Science Library Books

Each *Lots of Science Library Book* is made up of 16 inside pages, plus a front and back cover. All the covers to the *Lots of Science Library Books* are located at the front of this section. The covers are followed by the inside pages of the books.

How to Photocopy the *Lots of Science Library Books*

As part of their *Great Science Adventure,* your students will create *Lots of Science Library Books*. The *Lots of Science Library Books* are provided as consumable pages which may be cut out of the *Great Science Adventures* book at the line on the top of each page. If, however, you wish to make photocopies for your students, you can do so by following the instructions below.

To photocopy the inside pages of the *Lots of Science Library Books*:

1. Note that there is a "Star" above the line at the top of each *LSLB* sheet.

2. Locate the *LSLB* sheet that has a Star on it above page 16. Position this sheet on the glass of your photocopier so the side of the sheet which contains page 16 is facing down, and the Star above page 16 is in the left corner closest to you. Photocopy the page.

3. Turn the *LSLB* sheet over so that the side of the *LSLB* sheet containing page 6 is now face down. Position the sheet so the Star above page 6 is again in the left corner closest to you.

4. Insert the previously photocopied paper into the copier again, inserting it face down, with the Star at the end of the sheet that enters the copier last. Photocopy the page.

5. Repeat steps 1 through 4, above, for each *LSLB* sheet.

To photocopy the covers of the *Lots of Science Library Books*:

1. Insert "Cover Sheet A" in the photocopier with a Star positioned in the left corner closest to you, facing down. Photocopy the page.

2. Turn "Cover Sheet A" over so that the side you just photocopied is now facing you. Position the sheet so the Star is again in the left corner closest to you, facing down.

3. Insert the previously photocopied paper into the copier again, inserting it face down, with the Star entering the copier last. Photocopy the page.

4. Repeat steps 1 through 3, above, for "Cover Sheets" B, C, D, E, and F.

Note: The owner of this book has permission to photocopy the *Lots of Science Library Book* pages and covers for classroom use only.

How to assemble the *Lots of Science Library Books*

Once you have made the photocopies or cut the consumable pages out of this book, you are ready to assemble your *Lots of Science Library Books*. To do so, follow these instructions:

1. Cut each sheet, both covers and inside pages, on the solid lines.

2. Lay the inside pages on top of one another in this order: pages 2 and 15, pages 4 and 13, pages 6 and 11, pages 8 and 9.

3. Fold the stacked pages on the dotted line, with pages 8 and 9 facing each other.

4. Turn the pages over so that pages 1 and 16 are on top.

5. Place the appropriate cover pages on top of the inside pages, with the front cover facing up.

6. Staple on the dotted line in two places.

You now have completed *Lots of Science Library Books*.

How do insects use their mouths and antennae?  *Lots of Science Library Book #4*	**What are the characteristics of insects?**  *Lots of Science Library Book #3*

Who studies insects?  *Lots of Science Library Book #2*	**What are insects?**  *Lots of Science Library Book #1*

A

antennae smell taste touch *olfactory nerves (ol FAK to ree) *mandibles *maxillae (max IL ee) *labium (LAY bi um) *stylet	Explain how insects observe their world. Describe insect mouthparts.
exoskeleton molt head thorax abdomen *chitin (KYE tin)	Describe the characteristics of insects. Explain how insects grow.
nature classification arthropod insect *symmetry *vertebrates *invertebrates *taxonomy	What are insects? What factors have made insects successful?
collection naturalist *identification *entomology	Describe a naturalist.

What is the form and function of insects' legs and wings?

Lots of Science Library Book #6

How do insects' respiratory and circulatory systems function?

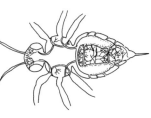

Lots of Science Library Book #8

How do insects see?

Lots of Science Library Book #5

How do insects reproduce?

Lots of Science Library Book #7

B

Explain how insects breathe.

How does "blood" move inside an insect's body?

breathe
closed pumps
oxygen
blood
*trachea (TRAY kee ah)
*spiracles (SPY rah kulz)
*hemolymph (HEE muh limf)

Describe insect legs and explain their uses.

Describe insect wings.

legs
wings
*coxa (KOX ah)
*femur (FEE mer)
*tibia (TIB ee ah)
*tarsus (TAR sus)
*prolegs
*halteres (hall TEERS)

Describe some courting methods of insects.

Explain how insects reproduce.

reproduction
egg
hatch
courtship
mating
scraper
file
*tympanum (tim PAN um)
*fertilize
*sperm

How do insects see?

Describe compound eyes.

vision
compound eyes
lenses
*facets
*lateral
*dorsal
*ultraviolet
*ocelli (o SEL ee)
*ommatidia (o mah TEEd ee ah)

How do insects defend themselves?

Lots of Science Library Book #12

What is incomplete metamorphosis?

Lots of Science Library Book #11

What is complete metamorphosis?

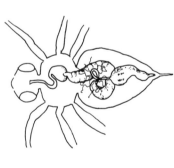

Lots of Science Library Book #10

How do insects' digestive and nervous systems function?

Lots of Science Library Book #9

Card 1

food
protect
prey
mimicry
camouflage

*defense
*offense

Explain camouflage and mimicry.

Describe other ways insects defend themselves.

Card 2

undeveloped
gills
color

*naiad (NYE ad)
*instar

Name and describe the stages of incomplete metamorphosis.

Compare and contrast dragonflies and damselflies.

Card 3

metamorphosis
(met ah MORE fuh sis)
egg
larva
pupa
adult

*imago (im MAH go)
*pupate

Name and describe the stages of complete metamorphosis.

Explain the advantages and disadvantages of complete metamorphosis.

Name some insects that pass through complete metamorphosis.

Card 4

food
digestion
tube
nerve

*ganglia (GANG lee ah)

Explain how insects digest their food.

Describe an insect's nervous system.

Why are ants and termites called social insects?

Lots of Science Library Book #16

Why do some insects migrate?

Lots of Science Library Book #15

Where do insects live?

Lots of Science Library Book #14

What do insects eat?

Lots of Science Library Book #13

D

---

colony
social insects
ants
termites
mature
reproduce
*reproductives
*soldiers
*workers
*sterile

What are social insects?
Describe an ant colony.
Describe a termite colony.

---

habitat
nest
water
food
shelter
*microhabitat
*environment

Where do insects live?
Name some insects and describe their habitat.

---

weather
cold
migration
*overwinter
*emigration

Explain migration.
Describe the migratory behavior of monarch butterflies.

---

predators
lure
wait
trap
*rotate

Describe some ways insects capture prey.
Name some predatory insects.

Are insects helpful or harmful?

Lots of Science Library Book #20

What are beetles?

Lots of Science Library Book #19

What are the differences between butterflies and moths?

Lots of Science Library Book #18

Why are honey bees called social insects?

Lots of Science Library Book #17

harmful
beneficial
predatory
pests
crops
*pesticides

Describe how insects are helpful.

Describe how insects are harmful.

---

beetles
successful
*coleoptera (kole ee OP ter ah)
*elytra (ee LIE trah)

Describe the characteristics of beetles.

Name some beetles.

---

butterflies
moths
caterpillar
chrysalis
cocoon
scales
*prolegs
*spinneret
*proboscis (proh BOS is)

Compare and contrast butterflies and moths.

---

honey bee
hive
honey
stingers
venom
*swarm
*pollen
*drones

Name and describe the members of a honeybee colony.

Explain how honey is produced.

What are mites and ticks? *Lots of Science Library Book #24*	What are spiders?  *Lots of Science Library Book #23*
	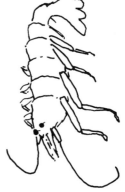
What are arachnids? 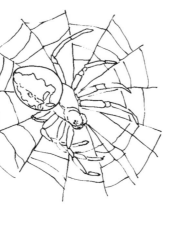 *Lots of Science Library Book #22*	What are crustaceans?  *Lots of Science Library Book #21*

mite
tick
blood
oval tear
*scutum (SKEW tum)
*capitulum (kah PITCH u lum)
*host
*questing
*asthma

spider web
oil
nest
*ballooning
*spiderlings

Compare and contrast hard and soft ticks.

Explain the various life cycles of hard ticks.

Name at least one disease transmitted by ticks.

Describe the characteristics of spiders.

Explain how spiders travel distances.

How do spiders catch their prey?

---

arachnids
liquid
*book lungs
*chelicerae (key LISS er ee)
*pedipalps (ped ee palps)
*carnivores

crustaceans
crab
lobster
krill
*cephalothorax (SEH fa lo thor ax)
*chelae (KEE lee)

Describe the characteristics of arachnids.

Name some arachnids.

Describe a crustacean.

Where do crustaceans live?

Name some crustaceans.

Page 5

Think about your mailing address. The following information helps in speedy mail delivery:

Country – USA
State – Florida
City – Melrose
Street Name – Main Street
House Number – 123
Last Name – Barnes
First Name – Terry

Terry Barnes
123 Main Street
Melrose, Florida, USA

Page 7

Taxonomy is the science of classifying living things. Carl Linnaeus (1707-1778) is considered the father of taxonomy. Although his original system of naming and classifying living things has gone through many changes, it is still used today.

Page 12

About 80% of all known animals are insects.

Scientists have classified more than 800,000 different species of insects, but believe this to be less than half the total number on Earth.

Page 10

Arthropods consist of five main classes:

1) crustaceans – crabs, lobsters, shrimp, crayfish, sowbugs, barnacles

2) arachnids – spiders, scorpions, ticks

Page 1

The animal kingdom can be divided into two types of animals: vertebrates and invertebrates.

Page 3

backbone

backbone

Page 16

Incredible Insights into Insects

The longest insect in the world is the giant stick insect of Indonesia. It grows to about 13 inches long, about the length of two new pencils.

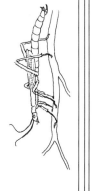

Page 14

Another reason for insects' success in survival is that they eat almost anything. Insects feed on meat, vegetables, fruit, blood, wood, waste products, and juices of flowers. Insects also have an extremely high rate of reproduction.

In a similar manner, scientists use a system of classification to identify animals.

    Kingdom – Animal
    Phylum – Arthropod
    Class – Insect
    Order – Hymenoptera
    Family – Apidae
    Genus – Apis
    Species - Mellifera

Vertebrates are animals with a backbone, such as mammals, birds, reptiles, fish, and amphibians. Invertebrates are animals without a backbone, from one-celled organisms to giant squids that reach 60 feet long.

Invertebrates make up about 96% of the animal kingdom.

Arthropods make up the largest phylum group in the animal kingdom. The word *arthropod* means "jointed limbs." The main characteristics of arthropods are:

1) an outside skeleton called an exoskeleton
2) legs and other parts attached to the body are jointed and can bend
3) bodies are segmented
4) one side of the body is a mirror image of the other side

3) diplopods – millipedes

4) chilopods - centipedes

5) insects – ants, butterflies, bees, beetles, cockroaches

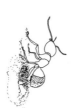

Animals are classified into groups by their similarities and differences. By classifying animals, scientists can study and understand their structure and behavior.

Insects adjust easily to their environment because of their size. They are so small that they can live and feed on resources that are too small for other animals. A caterpillar probably eats, during its lifetime, the equivalent of what a cow eats in one bite.

Pine trees produce a semi-solid substance called resin. Insects are sometimes trapped in this gummy substance. Fossilized resin is called amber. Insects have been found perfectly preserved in amber. Studies of fossilized insects show that today's insects have very similar characteristics to those that lived in prehistoric times.

During the 19th century, people became interested in nature and showcased their collections of insects and plants. People who study nature are called naturalists.

1

Naturalists need only a few simple tools, such as a field guide, magnifying glass, tweezers, bug box, journal, and pencil. A camera brings added enjoyment but is optional. There is no need to travel to exotic places because nature is in your own backyard.

3

A naturalist explores nature by observing, listening, recording, and sketching what is seen. When observing nature, keep these six sensible rules in mind:

5

4) Handle insects carefully so you do not harm them.
5) Do not keep insects captive for more than 24 hours. Release them in their natural habitat.
6) Do not litter. Be respectful of the natural world. Leave behind only your footprints.

7

1) *Blattodea* means "cockroach." This order, of course, includes cockroaches.
2) *Coleoptera* means "sheath wings." This order includes beetles, the largest order of insects.

10

5) *Ephemeroptera* means "short-lived wings." This order includes mayflies
6) *Hemiptera* means "half wings." This order includes chinch bugs, water bugs and bedbugs. These are considered "true bugs."

12

9) *Lepidoptera* means "scale wings." This order includes butterflies and moths.
10) *Mantodea* means "soothsayer." This order includes praying mantises.

14

Insects are everywhere. All the insects in the world combined would weigh more than all the humans combined.

Scientists sometimes have different opinions on the classification of arthropods, so bear this in mind when researching.

16

1) Always explore with a partner.
2) Let an adult know where you are and always ask permission before entering someone's private property.
3) Never touch a plant or insect if you are unsure of its identification. Some plants and insects are poisonous. Insects may bite or sting you.

---

3) *Dermaptera* means "skin wings." This order includes earwigs.

4) *Diptera* means "two wings." This order includes flies, gnats, midges, and mosquitoes.

Jean Henri Fabre (1823-1915), a French naturalist, wrote many books about insects. He studied insects by observing them in their natural habitats.

---

11) *Odonata* means "tooth jawed." This order include damselflies and dragonflies.

12) *Orthoptera* means "straight wings." This order includes grasshoppers and crickets.

The study of insects is called entomology, and scientists who study insects are called entomologists.

There are about 31 orders recognized by entomologists, with 12 main orders.

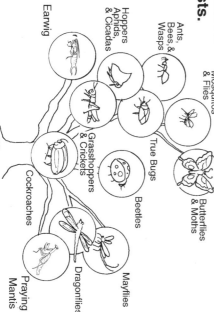

Earwig
Hoppers, Aphids, & Cicadas
Ants, Bees, & Wasps
Mosquitos & Flies
Grasshoppers & Crickets
True Bugs
Beetles
Butterflies & Moths
Cockroaches
Dragonflies
Mayflies
Praying Mantis

When you find an insect to observe, think about these questions: Where was it found? What was it feeding on? What time of day was it seen?

mouth parts

---

7) *Homoptera* means "similar wings." This order includes cicadas, aphids, and leafhoppers.

8) *Hymenoptera* means "membrane wings." This order includes bees, wasps, and ants.

Exoskeletons cover an insect's entire body – even the feet, eyes, and antennae. However, an exoskeleton does not grow with an insect. Therefore, an insect must molt, or shed, its exoskeleton several times during its lifetime as its size changes.

5

Incredible Insights into Insects

The atlas moth of India is one of the largest moths in the world. Its wingspan measures about eight inches across.

7

Did you know that not all insects are bugs? Bugs are insects of the Heteroptera or Hemiptera order, such as chinch bugs and bedbugs. True bugs have front wings that cover the hindwings. So not all insects are bugs, but all bugs are insects.

12

An insect's body is made of three parts: head, thorax, and abdomen. Specialized eyes, antennae, and mouth parts are located on an insect's head.

10

An evening walk in the rain forests of Ecuador may provide you with an impressive view of green and orange flashing lights. This light show is produced by a large click beetle that is the size of your finger.

1

Like all arthropods, insects have an outside skeleton called an exoskeleton. *Exo* comes from the Greek word meaning "outside." The exoskeleton is lightweight but strong. It acts like a suit of armor.

The exoskeleton is made of a tough material called chitin.

3

Fireflies, or lightning bugs, are neither true bugs nor true flies; they are a kind of beetle. The end section on their abdomen flashes a greenish light.

They use their light to send messages to potential mates. Each species has a unique code.

16

Incredible Insights into Insects

Fleas win the blue ribbon for high jumping. They measure only about 1/10 inch long but can jump about 8 inches. That's one hundred times their length!

14

Page 6

As a young insect grows, a new, soft skeleton forms underneath its exoskeleton. Eventually, the exoskeleton splits and the insect crawls out, complete with its new exoskeleton.

Page 8

That is equivalent to a human carrying a bulldozer.

Page 11

An insect's thorax is attached to its head, and six legs are attached to its thorax. Most adult insects also have one or two pairs of wings attached to the thorax, but some insects have no wings at all. Some species of insects only have wings during a certain stage of life. In other species, only one sex has wings.

Page 9

Insects are small, but can be amazingly strong. A leaf-cutter ant can carry an object 30 times its own weight.

A bumblebee can carry more than its weight in nectar and pollen. In comparison, a jumbo jet can carry only about 40% of its weight in cargo and passengers.

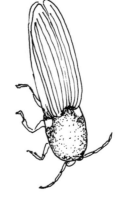

Page 2

This amazing click beetle is only one of the more than 800,000 species of insects that inhabit our world.

Page 4

Just as a tank has a hard covering to protect the inside, an insect's exoskeleton protects its soft body tissues from potential dangers.

Page 15

An insect's abdomen is attached to its thorax. The abdomen houses organs used for elimination, digestion, and reproduction.

Many female insects have an ovipositor at the end of their abdomen, which is used to lay eggs.

Page 13

Some insects' legs, such as praying mantises, are suited for grasping. Fleas and grasshoppers have strong legs adapted for jumping. Backswimmers' legs are adapted for swimming.

An insect's antennae are made of three segments. They contain olfactory nerves to help detect odors. Female moths give off a scent that makes it possible for male moths to detect them from several miles away.

The three segments of the antennae are the scape, pedicel, and flagellum.

5

Soldier and worker termites are blind and depend on their antennae to "see." Their antennae help them follow each other and communicate danger. Insects' antennae are so important that they cannot survive long without them.

7

Insects such as chinch bugs, water bugs, aphids, and cicadas have developed one mouthpart that pierces a plant or animal and another part that sucks fluid. These insects belong to the order Hemiptera, and they are true bugs.

Lots of Science Library Book #4   12

An insect's mouthparts vary greatly, depending on its diet. A female mosquito has a long, sharp, needle-like mouth for piercing the skin of animals and sucking their blood.

The female mosquito's piercing mouthpart is called a stylet.

10   Lots of Science Library Book #4

Insect antennae are used for smelling, tasting, and touching the world. These important sense organs are located on an insect's head between the eyes.

1

Lots of Science Library Book #4

Here are the different types of antennae.

aristate   capitate   clavate   lamellate

filliform   geniculate   pectinate

monoiliform   plumose   stylate

setaceous   serrate

Antennae may be simple or ornate.

3

★

Incredible Insights into Insects

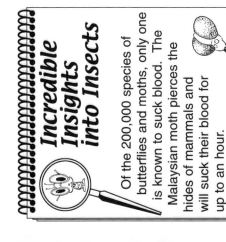

Of the 200,000 species of butterflies and moths, only one is known to suck blood. The Malaysian moth pierces the hides of mammals and will suck their blood for up to an hour.

16   Lots of Science Library Book #4

Wood-eating insects make up the largest group. Wood is not as nutritious a food as leafy plants, but it provides an abundant food supply.

14   Lots of Science Library Book #4

Some insects, such as bark beetles, give off a scent to attract other bark beetles. Mosquitoes rely on their antennae not only for smelling, but also for hearing.

Cave crickets use their antennae to sense vibrations. Scarab beetles have fan-shaped antennae that open up and detect smells and the wind's direction.

Most butterflies and moths do not have jaws, but a coiling tongue called a proboscis to siphon nectar. Their proboscis may extend an inch or more when feeding.

By observing the mouth of an insect, entomologists can determine its diet.

Mandibles are two strong, jaw-like mouthparts found in all chewing species. Maxillae are smaller jaws behind the mandibles. The labium is the lower lip. The labrum is the upper lip.

Most insects have two segmented antennae. Antennae of various insects look different. Some antennae are feathery, some are twisted, and some are cone-shaped. No matter how they are shaped, antennae are primarily used to smell and feel. Some insects use their antennae for tasting and hearing, too.

Incredible Insights into Insects

The longest antennae belong to longhorn beetles of Malaysia. Male longhorn beetles grow to about 3 inches long and their antennae measure about 9 inches.

The common housefly's lower lip has a sponge-like structure to absorb liquids.

The housefly's lower lip is called the labium.

Other plant-eating insects have cutting jaws. Some insects, such as grasshoppers, beetles, earwigs, and caterpillars, chew their food.

mouth parts

Insects have no eyelids, so their eyes are always open. Their eyes cannot move or focus as human eyes can.

There are two different types of ocelli: dorsal and lateral. Dorsal ocelli are located on the front of an insect's face. Most insects have three dorsal ocelli, usually located between the two compound eyes.

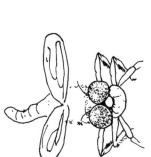

Most insects have eyes located on the sides of their heads. In various species, eyes may be located in other areas.

Only arthropods have compound eyes; specifically, insects and crustaceans.

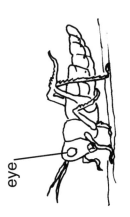

Many insects can see a wider range of light rays than humans can. Some insects, such as honeybees and butterflies, can see ultraviolet light, which is invisible to humans. Most insects see the colors yellow, green, blue, violet, and ultraviolet, but cannot see the color red.

Incredible Insights into Insects

The heaviest insect in the world is the Goliath beetle of Africa. It measures about 4 inches long and weighs about 3.5 ounces.

An insect's compound eye is made up of image producing parts called ommatidia.

An insect's sight is highly sensitive to movement. Insects can see shapes, and some insects have good depth perception as well.

Most insects in their immature, or larval stage, have lateral ocelli located on the sides of their head.

Many insects have eyes that are made up of many six-sided lenses, or facets. These eyes are called compound eyes because "compound" means two or more parts.

A compound eye may consist of a few to a few thousand lenses. A worker ant's eyes consist of about six lenses. The compound eyes of a dragonfly consist of more than 20,000 lenses, allowing it to see and catch fast-moving prey. Compound eyes are highly sensitive to movement.

The many lenses of compound eyes give the insect a mosaic of images that form a whole picture. How the insect interprets these images is studied by entomologists.

Not all insects have keen eyesight like the dragonfly. Some insects have simple eyes called ocelli. A simple eye, or ocellus, is made up of a single lens. Insects' ocelli distinguish light and dark but are unable to distinguish images.

Look at a rainbow. It contains a range of color that humans can see. Humans can see the colors red, orange, yellow, green, blue, indigo, and violet.

Fleas can leap more than a hundred times their body length. Their leg muscles push down on the rubbery pads on their feet and when they release the pressure, the pads send them leaping high in the air.

5

When an insect's wings are not in use, they lie flat. However, some insects fold their wings for protection.

12

Most insects have one or two pairs of wings which are attached to the thorax. Their wings have a network of veins. Butterflies and moths have wings made of overlapping scales.

10

7

All insects have three pairs of legs which are attached to the thorax. Insect legs are jointed, and each leg consists of four main parts: coxa, thigh, lower leg, and foot.

1

---

1) The coxa is the top of the leg, attached to the thorax.
2) The femur is the muscular part between the coxa and the lower leg.
3) The tibia is the lower part of the leg.
4) The tarsus is the foot, or tip of the leg, which usually consists of the claws and pad.

3

Some insects with two pairs of wings, such as beetles, use their front wings to protect their more delicate hind wings. The hind wings fold underneath the stronger front wings.

The front pair of wings are called elytra.

16

The wings of flies flap about 190 times per second. Honeybee wings flap about 250 times per second. Mosquito wings flap about 300 times per second.

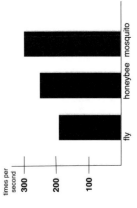

14

Imagine how you would walk with six legs. Insects move their front right leg, back right leg, and middle left leg at the same time. When those legs come down, they lift their front left leg, back left leg, and middle right leg.

Insect wings have no muscles, but the muscles inside the thorax move the wings up and down. The rate at which an insect's wing flaps varies considerably. Some butterflies have wings that flap only two times a second. Other insects, such as gnats, flap their wings more than 1,000 times a second.

Although not all insects fly, their ability to fly is one of the reasons for their survival. By flying, they can escape a predator and find new places to get food.

Insects are the only invertebrates, animals without a backbone, that have wings. Insects, birds, and bats are the only flying animals.

Insect legs may be used for walking, running, jumping, swimming, digging, catching prey, or collecting pollen. Common houseflies can walk on walls and ceilings. Special pads between their claws provide suction and enable them to walk upside-down.

Incredible Insights into Insects

The fastest flying insect is the deer bot-fly, which can reach speeds of up to 50 MPH. Dragonflies are a close second, reaching about 34 MPH.

Most flying insects have two pairs of wings, but some have one pair plus a second pair of wings that are used for balancing.

The modified wings used for balancing are called halteres.

Male California locusts jump several feet in the air, showing off their hind wings. As they jump, they beat their wings and make a loud, raspy sound. They land on the ground and kick their legs back and forth to impress the female.

5

Make crickets make this call to attract female crickets. Most female crickets do not chirp. Each species has different chirps that can be heard by other crickets of the same species.

7

Reproduction in most insects is sexual. Sexual reproduction occurs when a male insect fertilizes a female. A female lays the eggs.

The external sexual organ of a female insect is the ovipositor, used to lay the eggs in a specific spot.
A few insects give birth to live young.

12 Lots of Science Library Book #7

Flying male fireflies, or lightning bugs, send out flashes of light in a special sequence to attract wingless females. Females sit and wait on plants and respond by duplicating the flashes.

10 Lots of Science Library Book #7

When insects reach adulthood, their main priority is to find a mate for the purpose of reproduction. Insects use a variety of ways to attract mates, including smell, touch, sound, and special displays.

1 Lots of Science Library Book #7

During mating season, the female mourning cloak butterfly flies high in the air, followed by a male. Quickly, the female will make a dive downward, and if the male butterfly is able to keep up with her, mating will take place on the ground.

3

Insects lay their eggs in the ground, on water, on animals, in waste, and on plants. Insects lay eggs in areas that will provide food for the newly hatched larvae to consume.

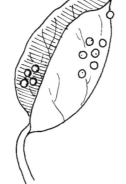

16 Lots of Science Library Book #7

Some insects do not go through internal fertilization. For example, a female insect may lay eggs and then a male fertilizes them where they were laid. Insect eggs come in a variety of shapes and sizes. Eggs may be round, flat, oval, or oblate.

14 Lots of Science Library Book #7

Male crickets make chirping sounds by rubbing their front wings together. The front wings have sharp edges, called scrapers, that rub on a bumpy ridge of vein, called a file, located under the wing.

Male burying beetles attract females by releasing pheromones onto a dead animal. Pheromones are chemical odors undetectable by animals, which signal mating or other behaviors. After mating, the beetles bury the carcass. Then the female lays her eggs beside it, providing food for the larvae when they hatch.

Insects that mate in swarms usually have a special locking structure on their bodies so they will not be separated. After a male honeybee mates with the queen, its mating part breaks off inside the queen.

During the summer months, you can hear cicadas, crickets, and katydids chirping and making loud noises. These are sounds of courtship. Male insects are calling out to attract females for mating.

Males of other butterfly species establish an airspace territory and keep other males away. They wait for a female butterfly to enter their airspace and then mate.

Insect eggs may be laid singly or in groups. Some insects, such as cockroaches, carry the eggs in a capsule. The capsule is called an ootheca. American cockroaches deposit the ootheca immediately; German cockroaches carry the ootheca, extended from the ovipositor, and deposit it when the eggs are ready to hatch.

In sexual reproduction, a male insect uses a pair of claspers to hold the female. Sperm, the reproductive cells, are passed into the female. The male sex organ used to pass the sperm is called the aedeagus.

air sac

air sac

5

The insect's circulatory system is important for many reasons. When an insect is injured, clotting occurs. The circulatory system helps destroy internal parasites or microorganisms that could harm the insect.

12

7

In a closed circulatory system, blood moves within vessels called veins and arteries. In an open circulatory system, an insect's blood, called hemolymph, flows freely throughout its entire body cavity.

10

All insects breathe through a network of tubes called a trachea.

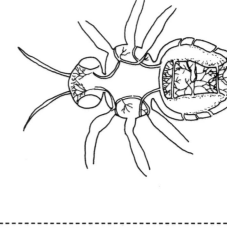

1

3

trachea

oxygen enters and carbon dioxide exits through spiracles

Incredible Insights into Insects

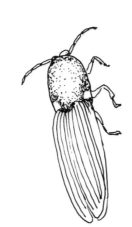

It's a good thing that the eggs of flies do not all hatch and survive. If one pair of flies mated and all the eggs hatched, lived to adulthood, and all their eggs hatched and survived, the number of flies would total almost 200,000,000,000,000,000. That is enough flies to cover Earth about 45 feet deep.

16

Human blood contains hemoglobin, which gives the blood a red color. Because insects lack hemoglobin, their blood is not red. Insect hemolymph is a watery fluid, usually clear or greenish. If you swat a mosquito on your arm and see red blood, it is probably your own.

14

These breathing tubes are connected to tiny outside openings on each side of an insect's abdomen and thorax. The trachea carries oxygen throughout the insect's body and carries carbon dioxide out. The openings on the sides of the abdomen and thorax are called spiracles.

2

Air enters the spiracles and is carried throughout an insect's body by the trachea. Active insects, such as grasshoppers, also have two large air sacs. These air sacs supply the flight muscles in the thorax with air.

4

A tube called a dorsal vessel runs along an insect's thorax and abdomen. The heart is within the dorsal vessel in the abdomen. As it beats, it moves the blood around the body.

In warm weather, an insect's heart pumps more quickly than in cooler weather. Why?

6

The muscular valves of the spiracles are closed most of the time, but they open to take in oxygen and release carbon dioxide. Air sacs increase the volume of air that insects take in and release.

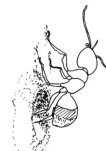

8

In insects of simple structure, the intake of oxygen and release of carbon dioxide takes place by diffusion, or the spreading and mixing of gas molecules throughout their body.

The insect has a heart, but it is not like a human heart. Insects have an open circulatory system, unlike the closed circulatory system of vertebrates.

9

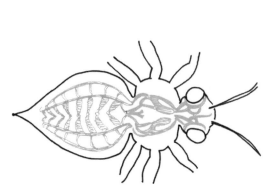

13

The digestive systems of insects are specialized to meet the need of their particular diets. Insects that eat solid food have gizzards to help them break down the food before it enters the midgut.

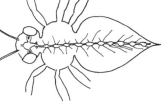

5

Insects have a well-developed nervous system. Nerve cords run parallel, or side by side, under their bodies from head to abdomen.

The nerve centers are called ganglia.

7

Most insects have a second nerve center in the head that controls the movement of their mouthparts.

12

Incredible Insights into Insects

Hairy winged beetles measure about 1/100 of an inch, small enough to crawl through the eye of a needle.

10

Most insects have a digestive system made up of a simple tube. The tube is divided into the foregut, midgut, and hindgut.

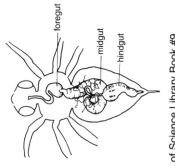

1

Digestion takes place primarily in the midgut. Food is absorbed in both the midgut and hindgut.

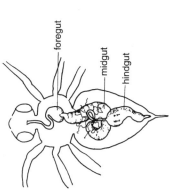

3

Since these nerve clusters work independently of the insect's brain, many insects can do ordinary tasks – such as moving about, laying eggs, and mating – without their heads.

16

In each thorax and abdomen segment, nerve clusters help control the movement of the segment in which it is located. They act like little brains, separate from the insect's true brain.

These nerve clusters are called ganglia.

14

Incredible Insights into Insects

Green-colored larva of polyphemus moths eat 86,000 times their birth weight in the first 48 hours of life.

There are two kinds of metamorphoses: complete metamorphosis and incomplete metamorphosis. Some wingless insects, such as silverfish, do not go through any kind of metamorphosis. Young silverfish look very similar to adult silverfish. They molt until they are fully grown.

Insects that go through complete metamorphosis have one major disadvantage: They cannot move during their pupa stage. They are vulnerable to parasites and predators. Therefore, larvae must find a safe place to become a pupa.

About 80% of all insects go through the four stages of complete metamorphosis: egg, larva, pupa, and adult.

egg
larva
pupa
adult

As a larva grows, it becomes too big for its exoskeleton, so it molts. The larva emerges, and soon its new exoskeleton hardens. The number of times a larva molts varies with each species.

The larval stages of some insects are so distinct that they have special names. For example, beetle larvae are called grubs. Fly larvae are called maggots. Butterfly and moth larvae are called caterpillars.

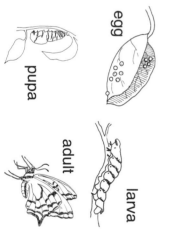

After a larva has eaten enough, it becomes a pupa. Pupa is a Latin word meaning "doll." A pupa looks like a doll wrapped in a blanket. During the pupa stage, insects do not move around or eat. Their bodies break down completely and reassemble inside the protective case.

An adult insect's job is to mate. Females mate and then lay eggs. The length of time an insect lives as an adult varies considerably. A mayfly emerges as an adult, mates, lays eggs, and dies in one day.

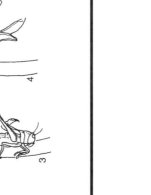

Another insect that undergoes incomplete metamorphosis is the damselfly. It spends a considerable amount of time in the nymph stage, molting about 12 times.

Not all insects undergo complete metamorphosis. Some insects change gradually as they grow and do not pass through the larval stage. This is called incomplete metamorphosis.

Dragonflies also undergo incomplete metamorphosis. Dragonflies are usually larger than damselflies. At rest, dragonflies hold their wings sideways; damselflies hold their wings together vertically.

Long-horned grasshoppers, more commonly known as katydids, are an example of an insect that undergoes incomplete metamorphosis. They are called katydids because their song sounds like "katy-did, katy-didn't."

A nymph is smaller than an adult, has undeveloped wings, and is a different color, often transparent. A nymph usually lives and eats in the same environment as an adult.

Young damselfly nymphs are often transparent and darken as they mature. They live underwater, absorb oxygen, and release carbon dioxide in a manner similar to that of fish.

Some insects take an unusual position or flare their wings to exaggerate their size. When threatened, a Chinese praying mantis opens its wings to make its body look bigger.

Stinkbugs, red-humped apple worms, and some caterpillars emit a strong, offensive odor. These smells keep some predators away.

Have you ever caught a butterfly, let it go, and looked at your fingers? Were your fingertips covered with colored scales? As butterflies and moths fly about, the chance of their running into a spider web is very high.

Insects have many other methods of defense. Long-horned beetles and katydids give a sharp bite when touched. Grasshoppers use their powerful legs to jump away from predators.

Hornets and bees can produce a painful sting when threatened. Some caterpillars are covered with hairs that release poison and are very irritating when touched.

Most bats eat nocturnal flying insects, like moths. Bats make high-frequency sounds that echo off their prey. Echolocation guides the bats as they hunt in the dark. Bats provide a natural means of insect control.

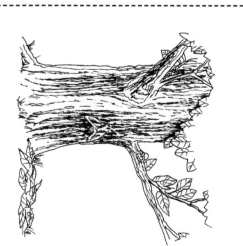

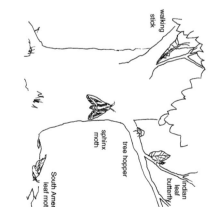

Rove beetles lure their prey. If dung or a carcass is not available, rove beetles secrete a strong-smelling substance and smear it on a leaf. The smell attracts small flies that land on the leaf. Then the rove beetle quickly snatches the flies and eats them.

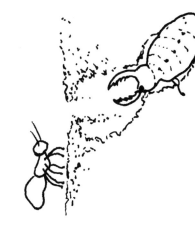

Praying mantises usually bite their prey with their powerful jaws. Praying mantises are the only insects that can move their heads from side to side. They can rotate their heads 180 degrees.

Praying mantises often sway gently back and forth, imitating the wind's movement. When the prey is close enough, praying mantises ambush it with their powerful front legs. Their front legs have sharp spines that prevent escape.

Most insects eat plants or plant products such as wood, paper, fabric, cork, or flour. Some insects hunt other animals for food. They are called predators. Predatory insects have a variety of ways to catch their prey.

Tiger beetles aggressively run along the ground, attacking insects. Tiger beetle larvae use their heads to dig holes and then drag their prey into the holes to eat them.

Dragonflies are attracted to areas where there is a great amount of insect activity, such as ponds. They dart back and forth, eating midges and other insects.

Various types of soils provide habitats for many insects. Ant lions live in sandy soil; soil rich in peat is home to springtails and beetle larvae.

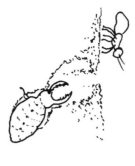

Insects also have habitats in water – not only ponds and lakes, but also hot springs and ice-cold mountain streams. About 5% of all insects spend the majority of their lives in water.

Some insects build nests. The females chew wood fibers and mix them with their saliva to build a protective, papery nest to hold their eggs.

Newly hatched wasps become workers, and they carry on the work to make the nest bigger. The entrance is small for temperature control, and it is usually located on the bottom of the nest.

An abundance of insects is found in tropical rain forests. The environment is the main factor in determining where insects are found in a region. That is why you will find more insects in a vacant, unmowed lot than in a neatly mowed yard. Fewer insects are found in very dry or very cold environments.

Animals can also provide habitats for insects. Fur, feathers, and skin provide habitats for fleas, lice, and other parasites.

Most aquatic insects live in water as larvae or nymphs and return to land as adults. Only a few insects, such as diving beetles and giant water bugs, live their entire lives in the water.

Some insects return to the surface for air, and other insects use air bubbles.

Page 5

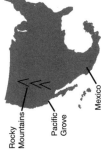

Page 7

Monarchs west of the Rocky Mountains migrate to Pacific Grove in California. Monarchs east of the mountain range migrate to central Mexico.

Page 12

Bogong moths travel from northern Australia to caves of southern Australia. They travel in great numbers and have an established migratory route.

Lots of Science Library Book #15

Page 10

Other butterflies and moths travel long distances, but they do not make a return trip.

This type of migration, called emigration, is primarily made to search for food.

Lots of Science Library Book #15

Page 1

Have you noticed that not as many insects are seen outside in cold weather? One reason is because many insects find hiding places to spend the winter.

Lots of Science Library Book #15

Page 3

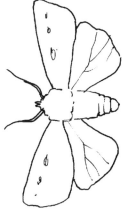

springtail

Page 16

To return to their nest, mud daubers use a combination of skills. They memorize landmarks, and they also utilize the Sun as a sort of compass or clock.

Lots of Science Library Book #15

Page 14

In Europe, Painted Ladies even migrate over the Alps. Due to their huge numbers and low flying level, they can cause problems like forced highway closings.

Lots of Science Library Book #15

Ants within a colony have specific roles called castes. Ant species differ somewhat, but usually there are thousands of female ants and one queen. The queen ant lays all the eggs.

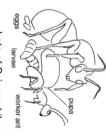
eggs / larvae / pupa / worker ant

Worker ants have various functions. Young worker ants generally take care of eggs, larvae, and pupae. Middle-aged worker ants dig tunnels and maintain the nest. Older worker ants gather food.

Sometimes larvae hatch with wings. These winged ants are both males and females. Worker ants give special care to these ant larvae by giving them extra food. When mature, the winged ants leave the nest to mate and begin new colonies. After mating, the new queen breaks off her wings.

The largest and most complex of all social insects are termites. A termite colony may consist of 100 to 1,000,000,000 termites. Within a colony are three types of termites: reproductives, soldiers, and workers.

The royal pair, male and female reproductives, are dark in color. Both the king and queen have wings and compound eyes. The queen lays all the eggs and the king fertilizes them. In some species, the king and queen live for more than ten years.

Termites in the tropics build elaborate termitaries that look like huge mounds. They are built out of soil and cemented together with saliva. They commonly reach about 20 feet high.

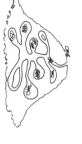

Rather than developing reproductive parts, workers develop stingers and venom. Workers take care of the queen, drones, and larvae by defending the hive with their stingers. Worker honeybees live about 30 days.

5

A queen lives about two years, considerably longer than drones and workers. Once the queen stops producing eggs, one of the fertilized eggs is moved to a special place and raised as the new queen.

Just before a new queen emerges from the pupa stage, the old queen and more than half the workers leave the nest together as a swarm to begin a new colony.

12

Another type of social insect is a honeybee. An average honeybee hive consists of about 20,000 bees. Honeybees are grouped into three types: drones, workers, and queen.

1

Workers fill the cells with pollen and nectar. They make the water in the nectar evaporate with their rapid wing movement. When the right amount of water has evaporated, the remaining mixture is honey. Workers then cap off each honeycomb cell.

16

Workers allow drones in the hive only during mating seasons, spring and summer. In autumn, workers keep drones out so they starve to death. No mating occurs during winter.

7

A queen produces a special chemical, called a pheromone. Pheromones are passed from worker to worker, causing them to continue working for the queen.

10

Because the sole purpose of drones is to mate with the queen, they lack the body parts to collect nectar and they have no stinger. Drones live about 20 days.

3

Swarming bees are generally docile, but still may sting. This man probably placed a queen on his head and waited for the workers to swarm around her. Caution: Do not try this.

14

If an enemy provokes a worker bee, it pierces the enemy with its stinger. The stinger tears from the worker's abdomen and it soon dies.

A queen lays fertilized and unfertilized eggs. Fertilized eggs turn into workers; unfertilized eggs turn into drones.

As with all social insects, the queen honeybee is the center of attention. A queen is the largest honeybee. She mates during only one time period in her life, and she stores the sperm cells from the several males in a special pouch in her body to use for the rest of her life.

Karl von Frisch (1886-1982), an Austrian scientist, studied honeybees and discovered how they communicate. He found that honeybees perform a circular dance when they locate food nearby. If the food is farther away, honeybees perform a more elaborate dance by wagging their abdomens.

Drones are male honeybees, born from the queen's unfertilized eggs. Within a hive, there are only a few hundred drones. Their only purpose is to mate with the queen. They mate in flight, and the drones die shortly afterward.

Workers are female honeybees that are sterile, or do not lay eggs. The workers are born from the queen's fertilized eggs and do all the work in the hive.

When a swarm of bees leaves a colony, they find a temporary place to cluster while some bees scout the area to find a permanent place. A swarm usually finds a permanent home within 24 hours.

Workers secrete a special wax from glands between the joints of their abdomen. They use the wax to build the cell walls of honeycombs. Honeycombs consist of six-sided compartments that hold honey and larvae.

A caterpillar's body is covered with sensory hairs called setae that give it the sense of touch.

When moth caterpillars are ready to pupate, they wrap themselves in silk threads. This is called a cocoon.

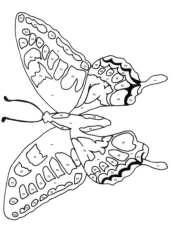

One difference between butterflies and moths is that butterflies are generally active during the day and rest at night, while moths are active during the night and rest during the day.

However, several moth species are specialized day flyers. Many of these moths are colorful and are often mistaken for butterflies.

Butterflies and moths differ from other insects in that their wings and bodies are covered with scales. They also have a long, coiled mouthpart used for sucking nectar.

Butterflies and moths go through complete metamorphosis. The four stages are egg, larva (caterpillar), pupa, and adult.

During this time, the caterpillar's job is to eat and eat and eat. Many caterpillars do not survive. They are susceptible to weather and have many predatory enemies.

Generally, butterflies hold their wings upright over their backs. Moths rest with their wings opened flat.

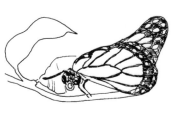

Generally, butterflies are more colorful than moths.

1 - black
2 - yellow
3 - yellow ocre
4 - orange
5 - red
6 - royal blue

Page 2

Caterpillars may appear to have a soft body, but they are covered with an exoskeleton, and they molt like other insects. A caterpillar has three body parts. It remains in this stage for about two weeks to a month, depending on the species.

Page 4

A caterpillar has twelve tiny eyes on its head. It has three pairs of legs on its thorax, used to hold the food it is eating. On its abdomen are four pairs of legs, called prolegs, used for walking and holding on to plants. A pair of claspers is located on the end of the caterpillar.

Page 6

When butterfly caterpillars enter the pupa stage, they attach to a plant and undergo a final molting. The caterpillar's skin hardens and becomes a pupa. A butterfly pupa is called a chrysalis.

Page 8

A Bombyx mori moth caterpillar secretes fluid through an opening in its head called the spinneret. As the fluid is exposed to the air, it hardens and forms a silk thread.

The silkworm spins a single thread, usually about 2,000 feet long, and wraps it around itself to form a cocoon. About 20,000 cocoons are needed to produce just one pound of silk cloth.

Page 9

Incredible Insights into Insects

Silkworms were originally found in China many years ago. The silkworm is no longer found in the wild, but only on silkworm farms. The silkworm is not a worm, but a Bombyx mori moth in its larval stage.

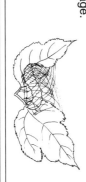

Page 11

The lifespan of butterflies and moths varies, from a few weeks to several years.

Page 13

There are many more species of moths than of butterflies.

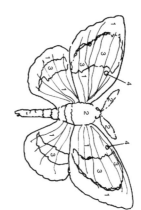

Page 15

Antennae of butterflies usually have a knobbed tip. Antennae of moths are usually feathery or plain.

1 - red brown
2 - dark
3 - beige
4 - white

Beetles are everywhere; they live in deserts, forests, mountains, and even hot springs. They are found on land, in the air, and in the water. Beetles vary greatly in size, from the minute featherwing beetle to the giant Goliath beetle. They eat all kinds of plants and both living and dead animals.

Beetles undergo complete metamorphosis: egg, larva, pupa, and adult.

egg pupa larva adult

Beetle larvae are called grubs. Some beetles remain in the grub stage for several years.

Beetles communicate by using pheromones, sounds, or visual signals. Tomentose Burying Beetles can use their antennae to locate a dead animal within a radius of two miles. The male and female beetle work together to bury the animal carcass. They mate and the female lays eggs on top of the burial mound so larvae can eat the carcass as soon as they hatch.

Some beetles are pests that feed on roots, leaves, and flowers of plants. They can do great damage to crops. For almost every type of food, there is a beetle that eats it. Some beetles are scavengers that live off of dead animals or animal waste.

The largest order of all living organisms is Coleoptera, commonly known as beetles. Coleoptera means "sheath wings." The word "beetle" comes from old English words meaning "little biter."

More than 60 families of beetles have been preserved in amber, or fossilized resin.

When beetles prepare for flight, they lift their elytra, allowing their hind wings to come out for flying.

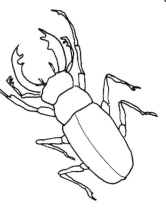

Incredible Insights into Insects

Lots of Science Library Book #19

The largest beetle in the world is found in South America. Titanus giganteus grows to about 20 cm long, large enough to fill a man's palm

---

Many beetles are black and appear plain, but others are beautiful with hues of metallic green, blue, and red. Natives of the Amazon have used the colorful and durable exoskeletons of beetles to make beautiful necklaces.

There are more kinds of beetles than any other kind of plant or animal. There are more than 350,000 species of beetles. One out of every five animal species is a beetle. By their numbers, beetles have shown to be one of the most prolific animals on Earth.

Often, beetles will open and close their elytra several times before taking off. By rapidly vibrating their elytra, they prepare their flight muscles for takeoff.

Elytra do not flap, but they provide lift for flying beetles.

Incredible Insights into Insects

Lots of Science Library Book #19

Male stag beetles have enormous antler-like mandibles and may inflict a painful bite, but the mandibles are used mainly as forceps to remove rivals from an area.

Some beetles, such as ladybugs, are considered beneficial to humans because as predatory insects, they kill other insects that damage crops. This is an example of biological control of insects by insects.

Incredible Insights into Insects

Lots of Science Library Book #19

The acteon beetle larva of South America is one of the largest insects in the world. The male can grow to 9 cm long, 5 cm wide, and 4 cm thick.

Beetles are clearly identified by their top wings, or elytra. The elytra is hard and is often shiny and colorful. The elytra protects the more delicate hind wings, which are folded underneath.

5

12

Although some insects are harmful to crops, there is usually another insect that eats the one causing the problem. Insects called cottony-cushion scales were destroying orange trees in California until ladybugs were brought in to eat them. The orange crop was saved.

Lots of Science Library Book #20

1

Although you may think of insects as harmful, annoying, and destructive creatures, about 99% of all insects cause no harm to humans. Insects keep nature in balance. They provide food for
mammals,
birds,
reptiles, fish,
and other
insects.

Lots of Science Library Book #20

16

Although termites are beneficial for cleaning the jungle floors, they can cause great damage in cities by eating the wooden structures of buildings.

Lots of Science Library Book #20

7

DDT also killed beneficial insects and entered the food chain, causing potential harm to other animals. The use of DDT was banned in the early 1970's.

DDT is the familiar name for dichlorodiphenyltrichloroethane.

10

Insects such as bees, butterflies, moths, flies, and beetles pollinate many plants. Honeybees not only collect pollen but also provide honey for humans and other animals.

Lots of Science Library Book #20

3

Some animals such as shrews, moles, hedgehogs, and anteaters, depend solely on insects for food. Insects are also a food source for some plants, such as the Venus flytrap.

Insect-eating animals are called insectivores.

Lots of Science Library Book #20

14

Harmful insects, or pests, make up only 1% of all insect species. Insects are considered pests if they cause injury to animals, or if they destroy food crops or structures. Mosquitoes can spread yellow fever and malaria. Malaria kills more than one million people every year.

Lots of Science Library Book #20

The benefits of insects are innumerable. Not only do delicate wings and colorful patterns add beauty to our world, but insects help us in other ways.

Silk is produced from the cocoons of silkworm moths. Natural dyes are made from insect secretions. Humans also eat insects. The early native Americans ate grasshoppers, as do people in parts of Africa today. Some people of Southeast Asia eat giant waterbugs made into a tasty condiment.

Keeping insects from eating crops is an ongoing problem for farmers. A chemical pesticide called DDT was designed to help with this problem. By 1950 it was used worldwide. Although it helped farmers considerably, the long-term effects were discovered to be harmful to the environment.

Scientists looked for new methods of pest control. Safer pesticides have since been created.

Biological control, which is the use of predatory and parasitic insects, has become more widely used today.

Many insects are predatory, which means that they eat other insects. Some predatory insects, such as ladybugs and praying mantises, are beneficial because they feed on pest insects and weeds.

Insect pheromones are used for pest control. By producing chemicals that mimic pheromones, scientists can bait insect traps.

They can also spray a field with the chemical to confuse the insects, thereby prohibiting reproduction.

Aphids weaken plants by sucking the sap. This eventually kills the plants. Because aphids reproduce without mating, they are prolific and very difficult to control.

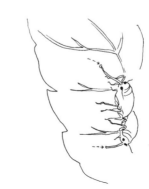

Another important function of insects is the breakdown of decaying matter and animal droppings. Scarab beetles, or dung beetles, lay their eggs in animal droppings. When the eggs hatch, larvae feed on the waste. Without dung beetles, we would have to be extra careful where we walk.

Crustaceans have two body parts. Crustaceans vary greatly in the number of appendages. They generally have five pairs of jointed appendages. One pair may be pincers.

The fused-together head and thorax are called the cephalothorax.

5

A crab is a crustacean. Crabs can grow quite large. The spider crab can measure 12 feet (3.66 m) across. This blue crab is one of several thousand species of crabs.

7

Krill are among the smallest crustaceans. They make up an important part of the food web. There are about 90 species of krill. Krill are small, shrimp-like crustaceans that swim in groups called swarms or clouds. Krill range in size from ½ inch to 3 inches (2.5 – 7.6 cm) long. They are a clear pinkish color. Some krill give off a bright, bluish-green light.

12   Lots of Science Library Book #21

Lobsters live on the dark ocean floor. Lobsters have compound eyes located at the end of stalks. Their long antennae are used to feel the environment, while the shorter antennae are used for smelling.

10   Lots of Science Library Book #21

The phylum arthropod consists of three main classes: insects, crustaceans, and arachnids. On land and in the air, insects are the most common class of arthropods. But in the water, crustaceans are just as plentiful.

1   Lots of Science Library Book #21

Most crustaceans are small, such as waterfleas, which are less than .039 inch (a millimeter) in length. A lobster, however, can grow to be about 24 inches (19.96 kilograms) long and can weigh about 44 lbs. (20 kilograms). There are about 39,000 species of crustaceans.

3

16   Lots of Science Library Book #21

Krill swim in large groups, turning the ocean pinkish in color. The most common krill is the Antarctic krill, found in the southern oceans. They live about 5 to 10 years. Scientists estimate that about 500 tons of krill live in the oceans.

14   Lots of Science Library Book #21

Most crustaceans reproduce sexually, meaning reproduction requires a male and a female. Some crustaceans, such as barnacles, possess both male and female reproductive parts.

The segments on a crustacean's body are called somites.

Lobsters continue to grow throughout their lives. Like other arthropods, they molt when they outgrow their exoskeleton.

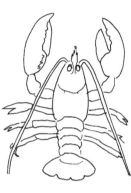

If a lobster loses one of its legs, it will eventually grow a new one. The process of growing a new part is called regeneration.

Crabs have flattened bodies that are almost entirely covered by their hard shell, or carapace. They have a small abdomen that is tucked under their body.

They have four pairs of walking legs and a pair of pincers called chelae. Crabs can walk sideways and burrow in the ground. They swim in water and live on land. Crabs have compound eyes located at the end of stalks, and they have good eyesight. They also have a keen sense of smell and taste.

Some saltwater crustaceans are shrimps, lobsters, crabs, barnacles, and krill. The crayfish is a freshwater crustacean. Wood lice, also called pillbugs, live on land in damp areas under stones and logs.

Krill eat one-celled plant organisms, called phytoplankton, that live in the surface water of the ocean. Krill live near the ocean floor away from their main predators, whales and birds. However, krill surface at night to feed.

Similar to other arthropods, crustaceans have an exoskeleton, jointed legs, and a symmetrical segmented body. However, crustaceans have two pairs of antennae and usually at least two pairs of mouthparts. Most crustaceans have a pair of compound eyes.

Arachnids have two body parts: anterior and posterior. The anterior consists of the head and thorax combined into one part. It holds the mouth, sense organs, and the appendages.

The anterior is called the cephalothorax.

5

Arachnids, like all arthropods, have an exoskeleton that they molt as they grow.

Arachnids do not have compound eyes. They have simple eyes, made up of just one lens.

7

The second pair of appendages is called pedipalps. Spiders' pedipalps serve as feelers.

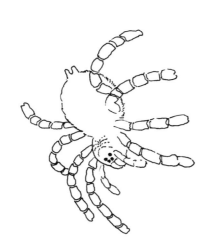

The use of pedipalps varies among species of arachnids. Pedipalps can serve as feelers, claws, pincers, or legs.

12 Lots of Science Library Book #22

Let's take a closer look at arachnids.

10 Lots of Science Library Book #22

Look at this spider. A quick glance at the spider makes you think that it is an insect.

1 Lots of Science Library Book #22

Arachnids and insects are similar in two ways - they have jointed legs and an exoskeleton.

arachnids

insect

3

Scorpions are carnivores, or flesh eaters, eating insects and small rodents. They paralyze their prey by injecting poison from the tip of their tail. The telson holds the venom gland. Most scorpion stings are somewhat painful, but not life-threatening.

16 Lots of Science Library Book #22

Mites make up the largest order of arachnids. Some mites feed on plants, causing serious damage to crops.

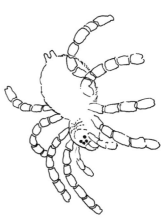

14 Lots of Science Library Book #22

Arachnids have several pairs of appendages. The first pair of appendages is called the chelicerae. On spiders, the chelicerae are fangs.

chelicerae

The other pairs of appendages, usually four, are true legs used for walking. Arachnids have no antennae and no wings.

Arachnids have mouthparts for biting but not chewing. Food is usually ingested as a liquid. Arachnids include scorpions, spiders, mites, and ticks.

There are about 60,000 known species of arachnids, and scientists believe there are many more yet to be identified.

The posterior is the abdomen, which consists of the reproductive parts. It also holds specialized breathing structures called book lungs. Small holes, called spiracles, on the body wall move oxygen into the book lungs and take carbon dioxide out.

Most arachnids live alone and only come together for mating. In most species, arachnids lay eggs. A few species bear live young. Scorpions give birth to a large litter of live young. They ride on the mother's back until they are ready to live independently.

Generally, arachnids are predatory animals. They hunt or lie in wait for small animals, often insects.

But take a closer look. Count the body parts. Count the legs. A spider has eight legs and only two body parts, so a spider is not an insect. The spider belongs to another class of arthropods called arachnids.

Most arachnids do not care for their young. The young must fend for themselves, and many die or are eaten by predators. Survival of the species is dependent on the large number of eggs laid or on how many young the arachnid bears.

After a spider has made its web, it waits. Unsuspecting prey get caught in the sticky web. The more the prey struggles, the more entangled it becomes. A spider does not get caught in its own web because it secretes special oil on its feet that allows it to move freely on the web.

5

Orb webs are beautiful displays of symmetrical designs. Other webs are triangle webs, tangle webs, and sheet webs.

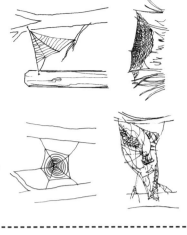

7

Spiders breathe with either two or four lungs. Air enters the lungs through spiracles on the body wall.

spiracles

12 Lots of Science Library Book #23

The most common spider webs are called orb webs. Spiders first spin the frame of the web. Then they spin silky thread out and around from the center, like the spokes of a wheel. Next, they spin thread to connect the spokes.

10 Lots of Science Library Book #23

The most familiar arachnids are spiders. Spiders have an exoskeleton and eight jointed legs. They have two body parts: The head and thorax are joined together to make one part, called the cephalothorax, and the abdomen makes up the second part.

1 Lots of Science Library Book #23

At the end of the abdomen, spiders have six openings, or spinnerets. Spiders secrete liquid from the spinnerets, and when the liquid is exposed to the air, it turns into a silky thread. Spiders use the silk threads to travel and build webs.

3

When spiders travel using a silky strand, it is called ballooning.

Some people think spiders look creepy, but they help us by eating insects that destroy crops. They also eat flies and mosquitoes that can spread disease. Most spiders are very beneficial.

16 Lots of Science Library Book #23

The female spider often eats the male after mating. After mating, the female spider lays eggs and spins a nest around the eggs. Some females die after laying the eggs.

The eggs hatch and the young stay inside the nest. In the spring, the spiders that have survived leave the nest.

14 Lots of Science Library Book #23

Although a spider has eyes, it cannot see well. Vibrations of the individual strands of a web inform the spider of its catch. Quickly, the spider wraps the prey with its silky thread, injects venom into the prey, and saves it for a future meal.

Spiders have simple eyes. Most spiders have eight eyes arranged in a definite pattern on their head, and in a different pattern for each species of spider. Spiders have fangs that inject poison. Most spiders eat insects, but a few eat frogs, fish, birds, and mice.

The spider's jaws are called chelicerae.

Spiders lay eggs. The spiders that hatch from the eggs look like the parents, only smaller. Spiders, like all arachnids, do not have wings.

Male wolf spiders assume a courtship posture when a female approaches. Their waving and drumming motions with their legs communicate their intentions to mate to the female.

All spiders spin thread, but not all spiders use it to build webs. Some spiders, like wolf spiders, trap-door spiders, and tarantulas, do not build webs to catch their prey.

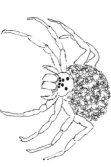

The female wolf spider wraps her egg sacs with silk thread and carries her developing young, or spiderlings, on her back for several weeks before they venture out on their own.

Incredible Insights into Insects

The silk strand from a spider's web is a strong natural fiber. It is five times stronger than a piece of steel the same size.

Incredible Insights into Insects

The world's largest spider is the theraphosa leblondi, a bird-eating spider of South America.

Tick larvae have six legs; nymphs and adults have eight legs. They have three body parts - head, thorax, and abdomen - that are fused together as one segment.

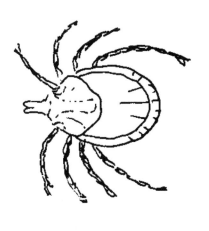

Ticks transmit a wide variety of diseases such as Lyme disease, Rocky Mountain spotted fever, relapsing fever, tularemia, and some forms of encephalitis.

Many hard ticks complete their life cycle on one host animal.

Mites make up the largest group of arachnids. There are about 30,000 species of mites. Mites are so small that they are often undetected.

mites on leaf

Ticks are divided into two groups: hard ticks and soft ticks. Hard ticks are found mainly on mammals, birds, and reptiles. Hard ticks are commonly teardrop shaped. Common hard ticks include the Lone Star Tick, Gulf Coast Tick, and American Dog Tick.

The family of hard ticks is called Ixodidae. Their body is covered with a hard shell called scutum.

Some mites live in the dust found in houses. They are called house dust mites and are a major cause of asthma among adults and children.

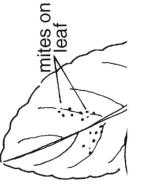

Hard ticks feed on their host for several days to several weeks, depending on the species. Their body expands to make room for the volume of blood ingested.

Ticks, a type of mite, are external parasitic arachnids. Parasites live off host animals, such as birds, mammals, and reptiles. A tick's diet is solely blood. There are about 850 species of ticks, and they are found worldwide.

The family of soft ticks is called Argasidae and lacks scutum.

Ticks lack a distinct head, although they have a pair of jaws attached directly to the body. Their piercing mouthparts consist of a pair of sharp mandibles and a pair of hooks, called rostrum.

Hard ticks have one blood meal during each of their developmental stages, larva, nymph, and adult. Female adult hard ticks reproduce once, laying thousands of eggs, and then die.

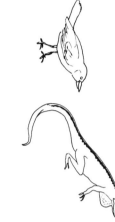

Adults
Female
Male
Nymph
Larva

0cm — 1cm

Other ticks live off one host during the larval and nymph stage, drop off, and then find another host on which to live their adult stage.

Commonly, ticks wait for a host to come to them. Ticks sense that a host is near by detecting carbon dioxide, heat, odors, and movement. When the host nears, they grab the host with their forelegs and quickly find a safe place to attach.

The behavior of ticks waiting on tall grass for a host, front legs extended, is called questing.

Most mites are terrestrial, or live on land, but some mites live in the water. Most mites are blood-sucking arachnids, but some feed on plants.

Soft ticks are less plentiful than hard ticks. They live primarily on birds, but are also found on bats and other animals. Soft ticks are usually oval shaped. The most common soft tick is the Spinose Ear Tick.

Great Science Adventures

Graphics Pages

Note: The owner of this book has permission to photocopy the *Graphics Pages* for classroom use only.

Investigative Loop™

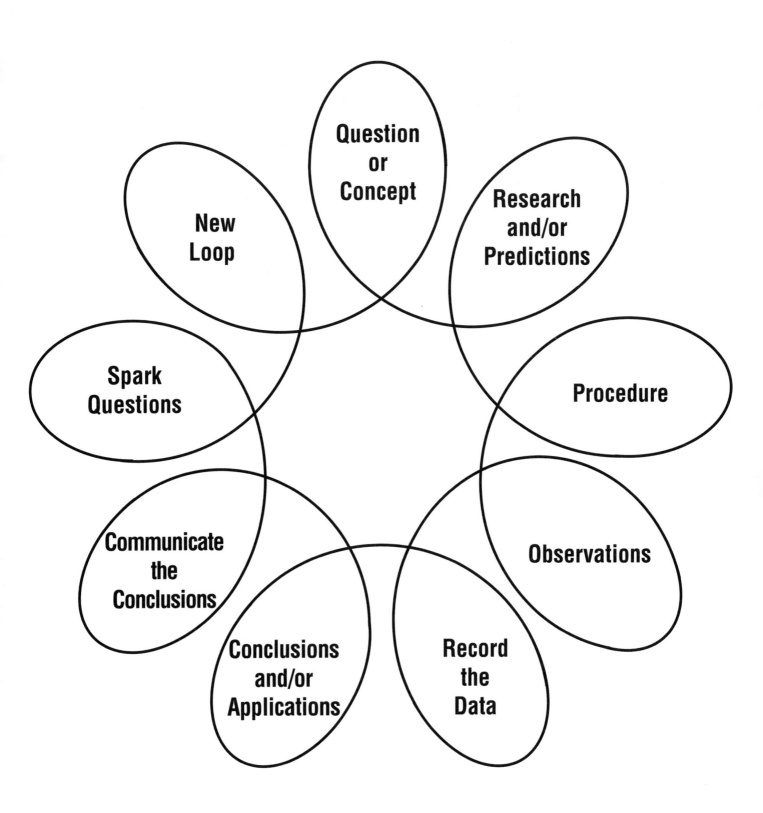

Problem Solving Scenarios

Problem: An African elephant has died of natural causes in a local zoo. It was an unusually large animal that lived a long and healthy life, well past average age expectations. Scientists want to preserve its skeleton for future study. How can they clean the skeleton of this huge animal so it can be reassembled and used for research? Hint: dermistid beetles.

Problem: Termites are known to be common within a region. A contractor who is building a new housing development wants to ensure home buyers that their homes will be termite free. Is this possible? What purpose do termites fulfill in nature? Investigate natural and chemical methods for controlling termites.

Problem: Museum officials note a fine dust appearing around certain items in their display cases. It is under and around baskets, woven natural fiber rugs, and items made out of animal hides. Upon closer examination, small empty cases, less than 1cm long, are found in the dust. What is happening? Is it a serious problem for the museum? What can be done about it?

Problem: A person is eating corn while dining in a restaurant and finds half a worm in it. The chef says it is not his fault because he was serving canned corn; the worm must have been in the can. Is it possible that the worm was in the can? What does the Food and Drug Administration say about the percentage of insect parts that can or cannot be in canned foods? Scientifically, is this a reasonable policy? Can insects be eaten as a source of protein?

Problem: Fruit flies are destroying cash crops in California. The state wants you to set up checkpoints to collect any fruit that tourists or business travelers might bring in from other states. Why? Is it possible to collect all imported fruit? What will you do with the collected fruit? Has a collection program like this been implemented before? If so, was it effective? Can other measures be taken? Evaluate the positive and negative effects of all proposals.

Classification 1A

| Classification | Kingdom | Phylum | Class |
| Order | Family | Genus | Species |

Insect Trap 1C

Symmetry 1D

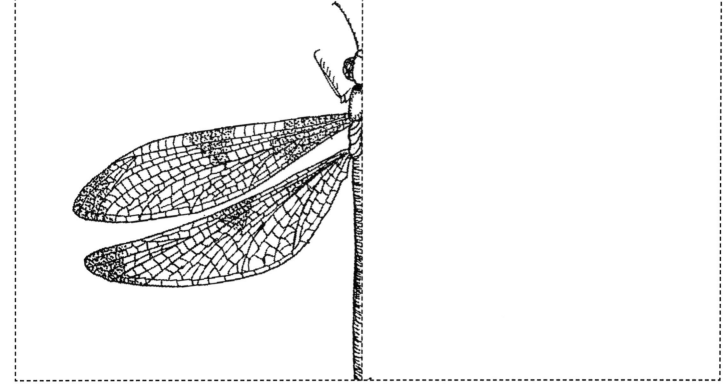

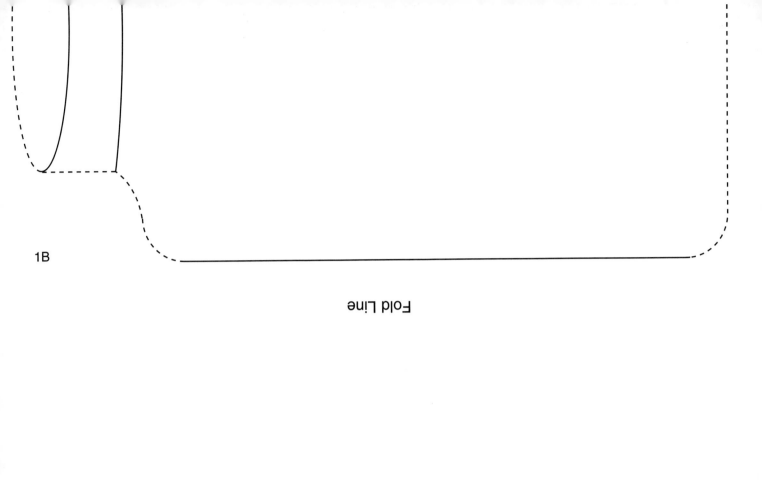
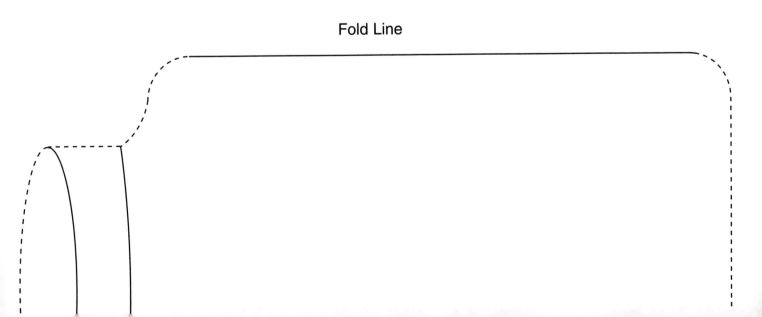

| Insect Collection | 2A | Catching Flying Insects | 2B |

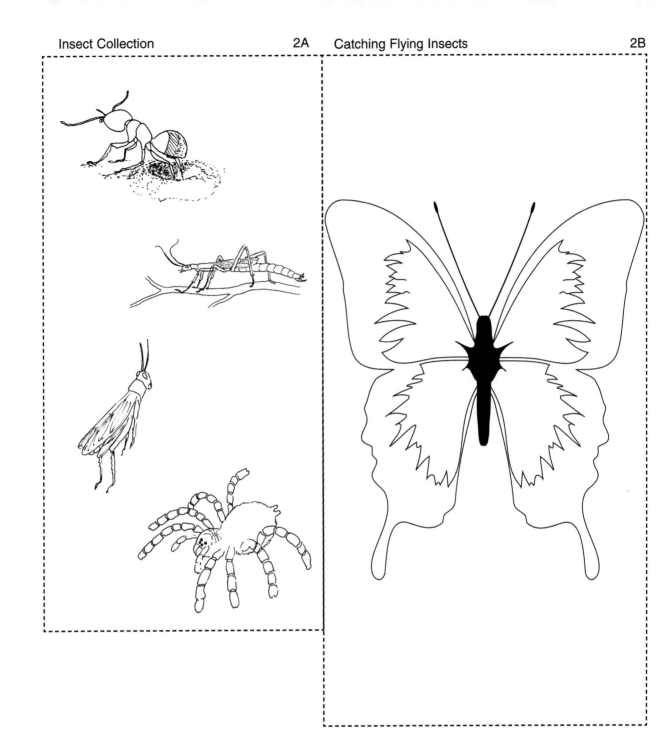

All About Insects

3A

3B

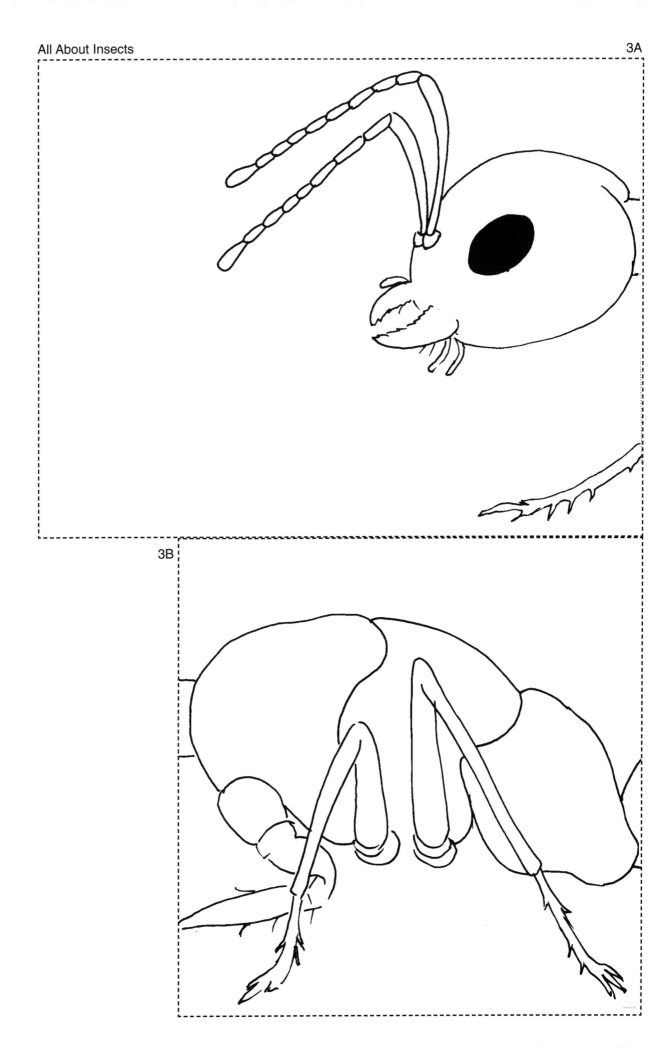

All About Insects 3C

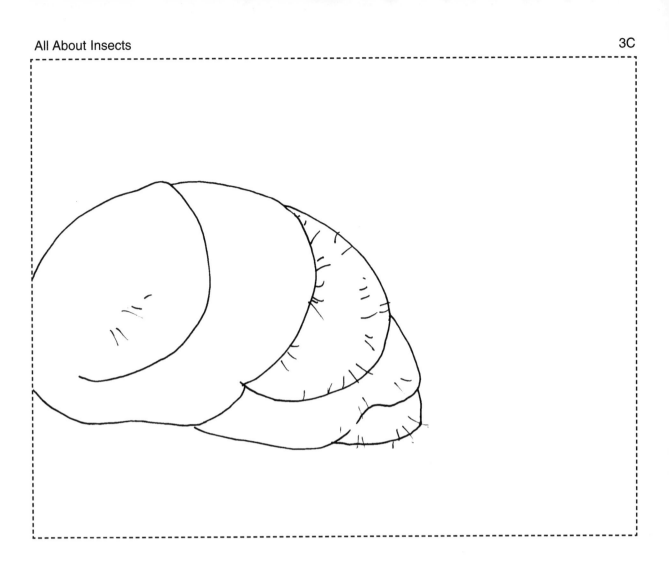

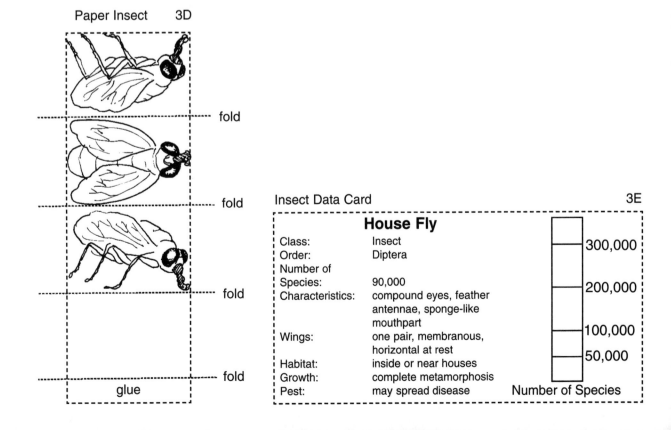

Paper Insect 3D

Insect Data Card 3E

House Fly

Class: Insect
Order: Diptera
Number of
Species: 90,000
Characteristics: compound eyes, feather antennae, sponge-like mouthpart
Wings: one pair, membranous, horizontal at rest
Habitat: inside or near houses
Growth: complete metamorphosis
Pest: may spread disease

Number of Species

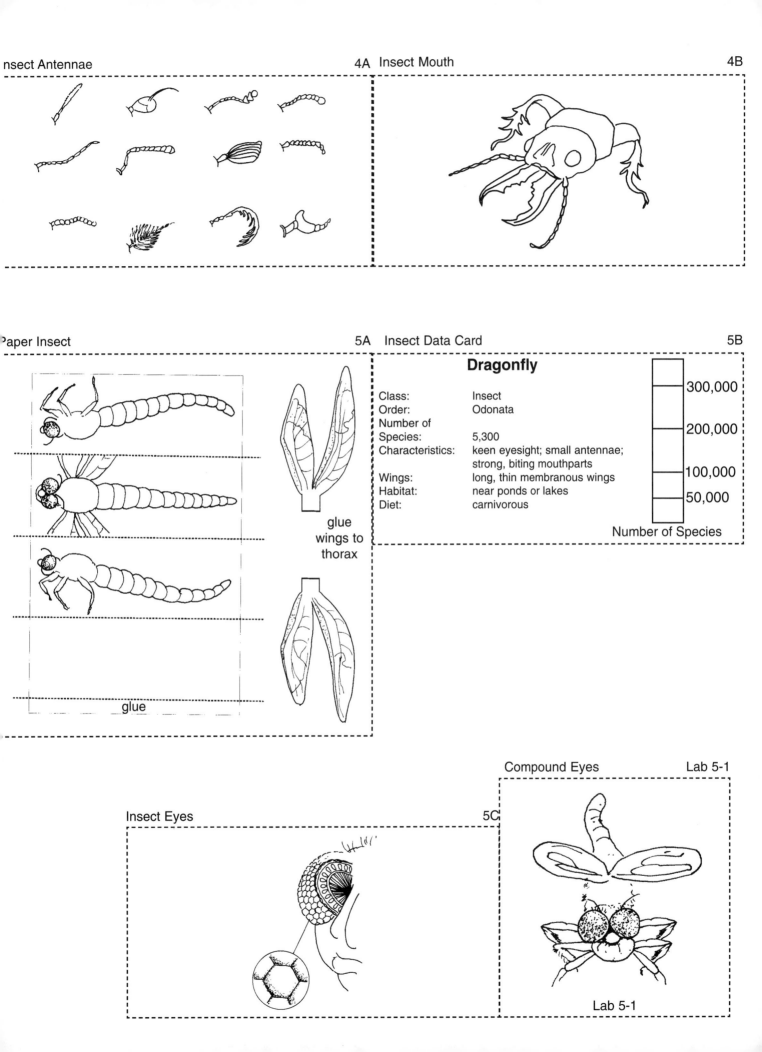

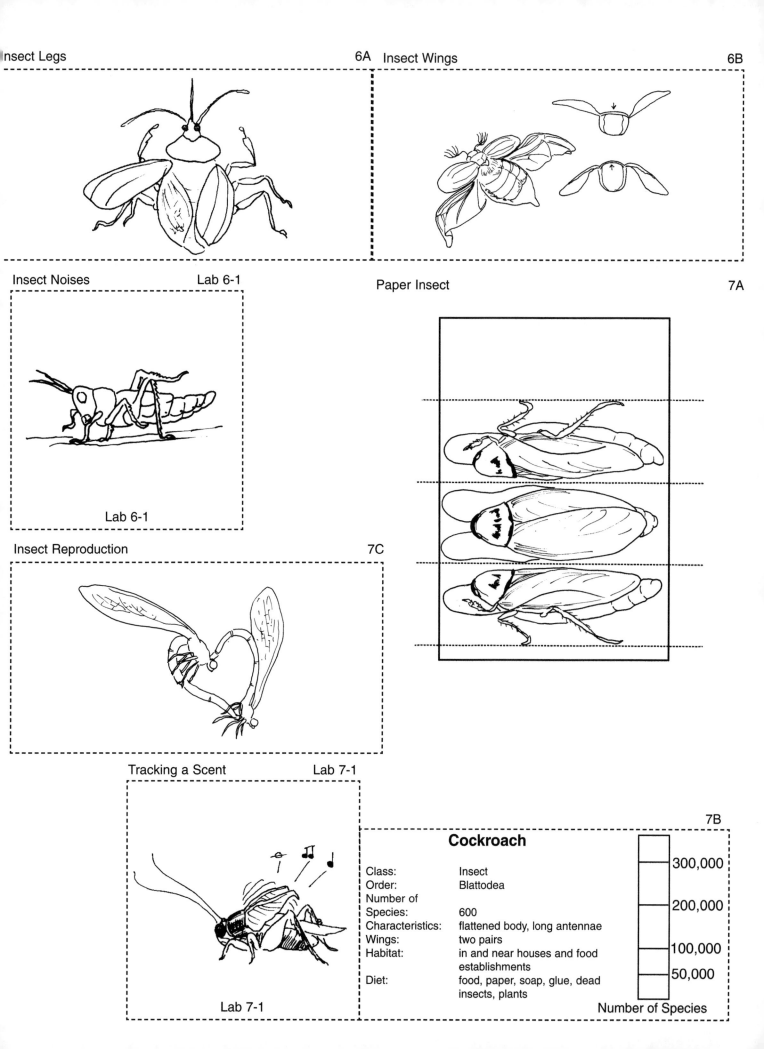

Insect Systems

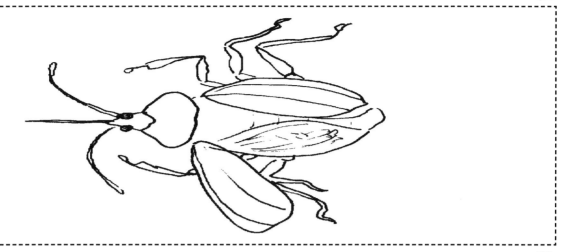

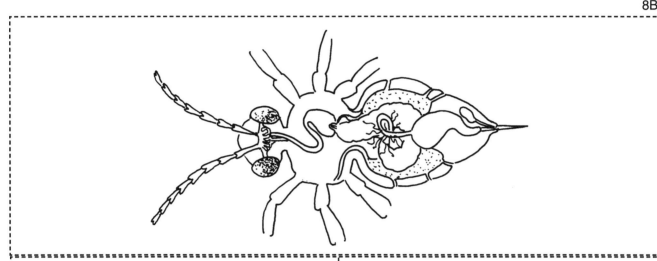

Respiratory System	Digestive System
Circulatory System	Nervous System

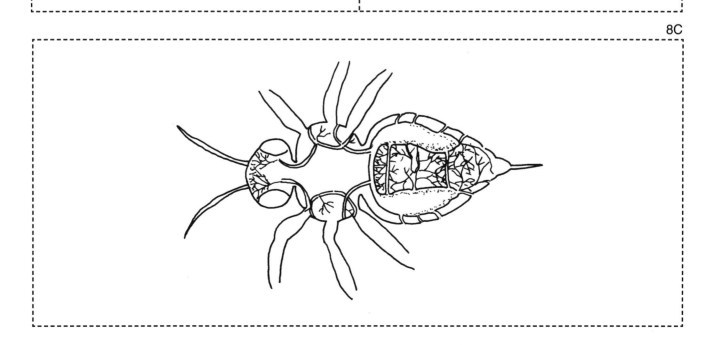

Insect Systems 8D

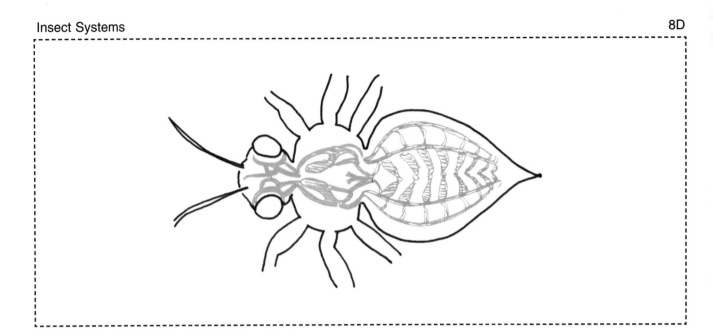

Insect's Heart — Lab 8-1

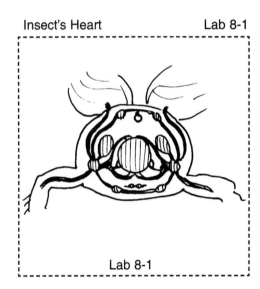

Lab 8-1

Insect Systems 9A&B

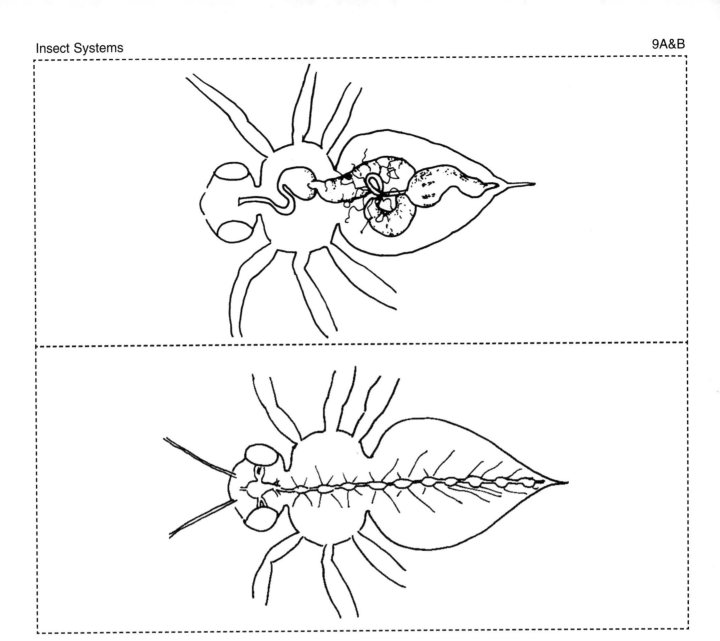

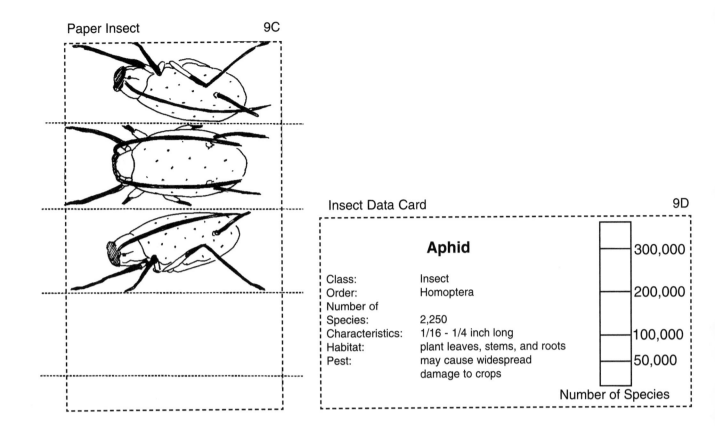
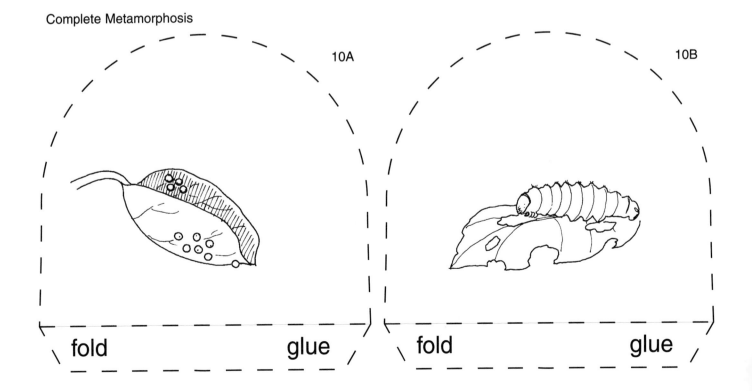

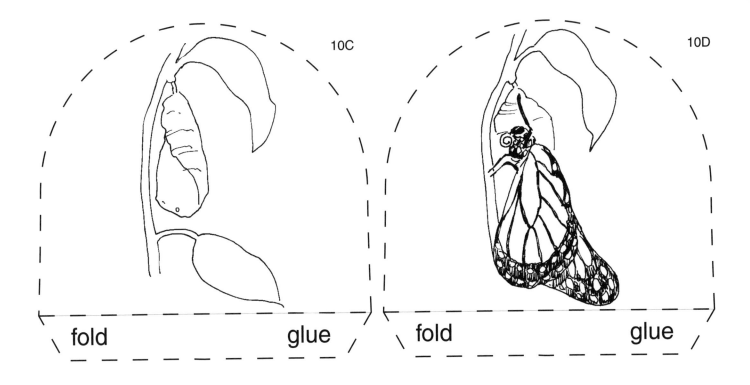

Butterfly and Chrysalis

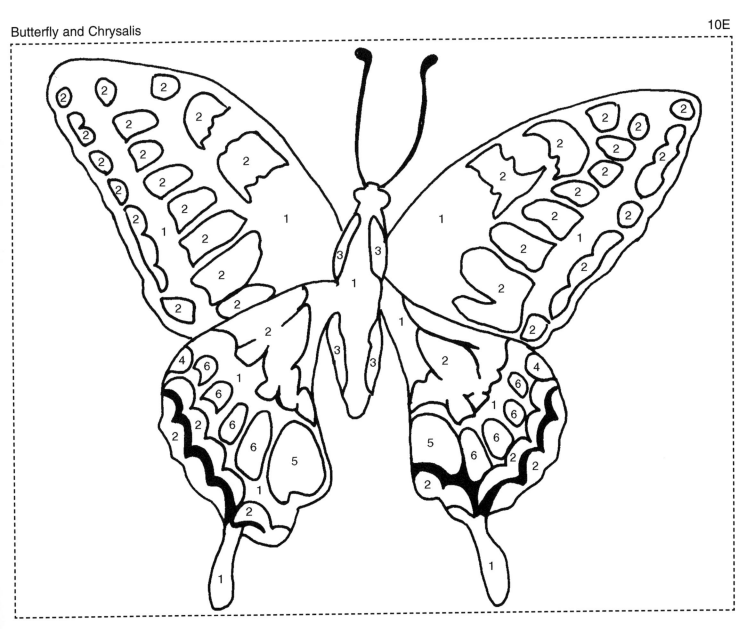

Incomplete Metamorphosis 11A	11B	11C

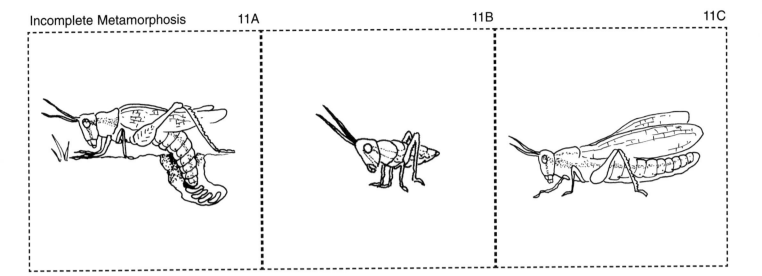

Insect Data Card 11E

Grasshopper

Class: Insect
Order: Orthopera
Number of Species: 20,000
Characteristics: 1-5 inches long, strong hind legs for jumping
Habitat: most commonly in lowland tropical forests semiarid regions, and grasslands
Diet: plants

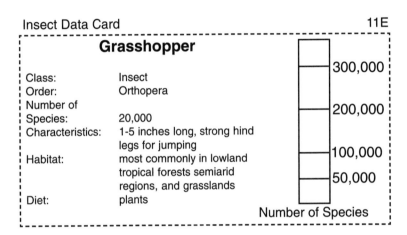

Number of Species

Paper Insect 11D

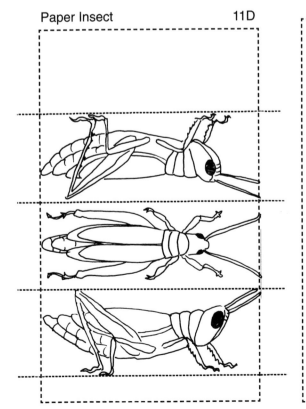

"The Two Voices" 11F

Today I saw the dragon-fly
Come from the wells where he did lie.
An inner impulse rent the veil
Of his old husk; from head to tail
Came out clear plates of sapphire mail.
He dried his wings; like gauze they grew;
Through crofts and pastures wet with dew
A living flash of light he flew.

by Alfred Lord Tennyson

Defenses 12A-H

Predatory Insects 13A

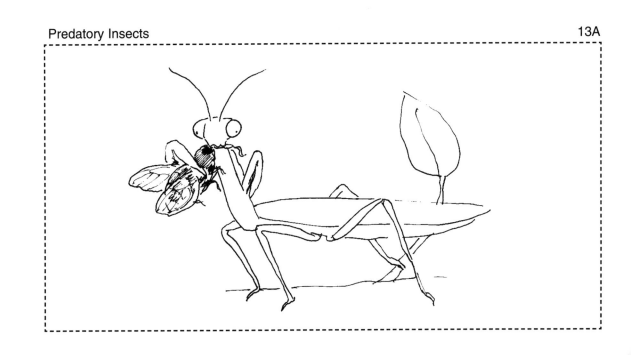

13B

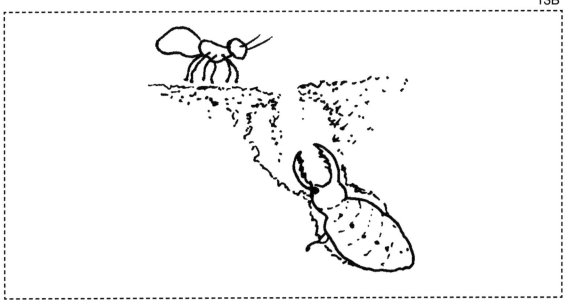

13C

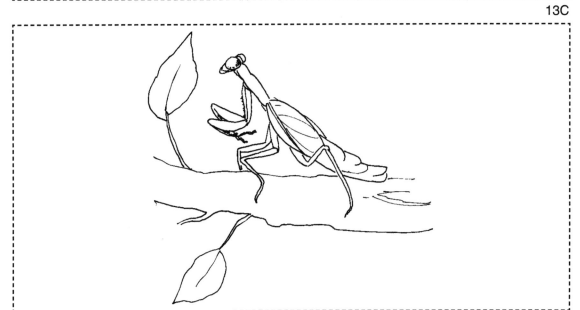

Predatory 13D

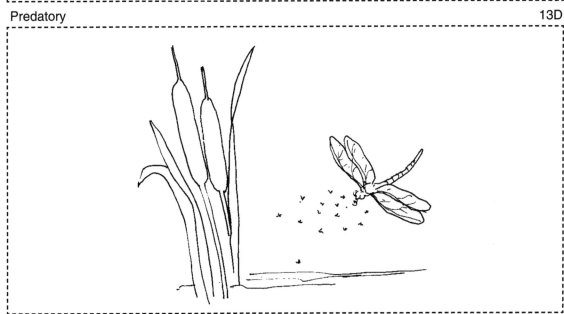

Paper Insects 13E

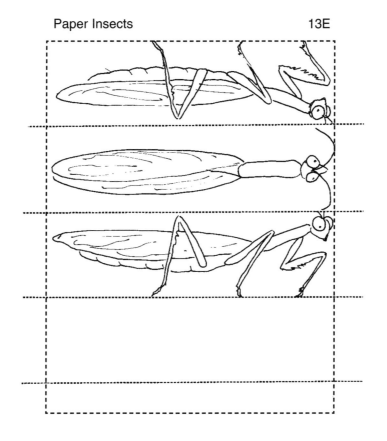

Insect Data Card 13F

Praying Mantis

Class:	Insect
Order:	Mantodea
Number of Species:	1,800
Characteristics:	up to 6 inches long, turns head up to 180 degrees, strong front legs
Habitat:	tropical and subtropical regions
Diet:	insects

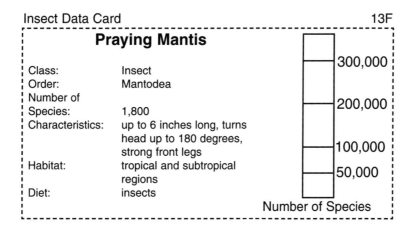

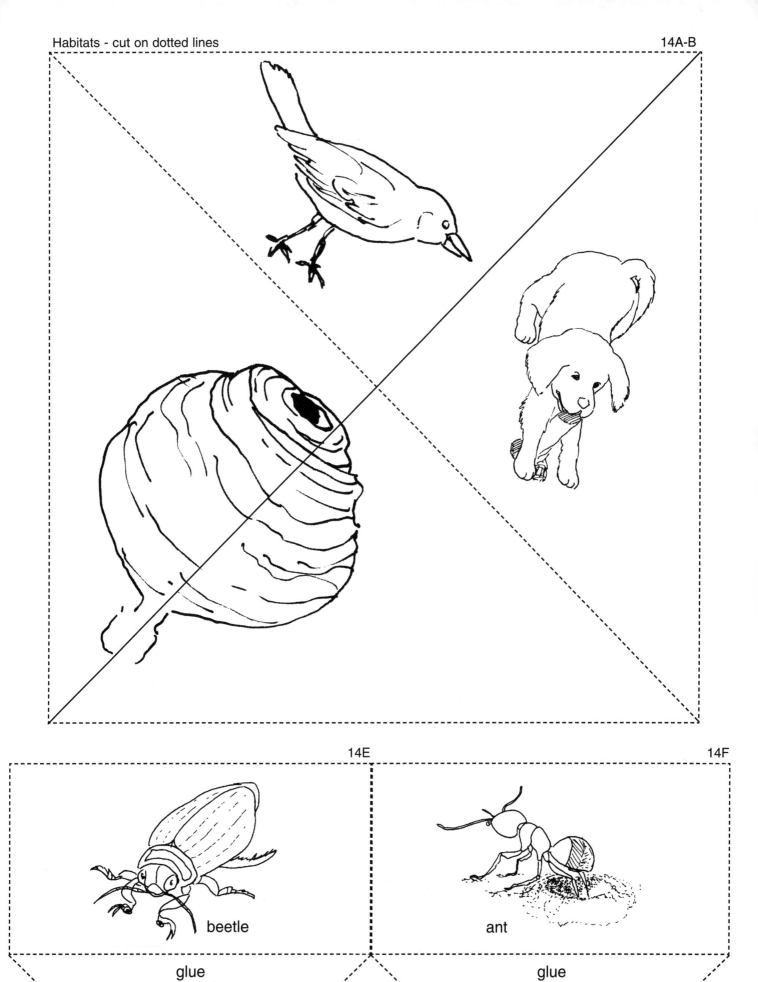

Habitats - cut on dotted lines   14C-D

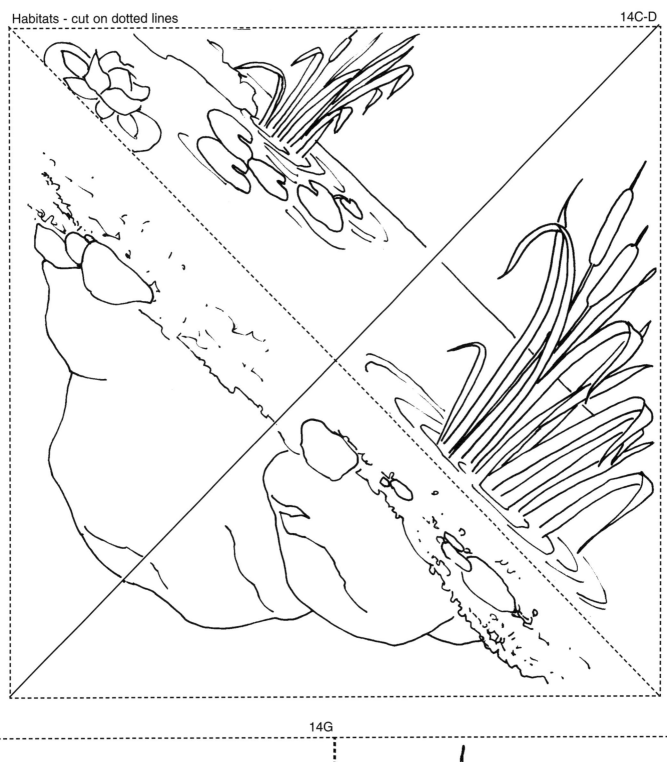

14G   14H

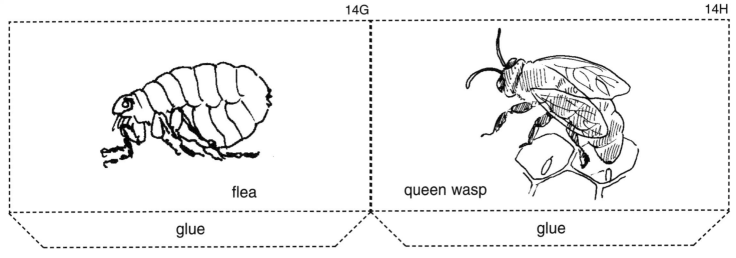

flea   queen wasp

glue   glue

Paper Insect 15A

Paper Insects 15B

glue wings on the thorax

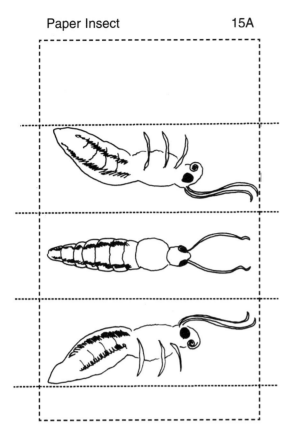

Insect Data Card 15C

Butterfly

Class: Insect
Order: Lepidoptera
Number of
Species: 18,500
Characteristics: proboscis mouthpart may be up to a foot long
Wings: two pairs
Habitat: woodlands, deserts, grasslands, tropics
Diet: flower nectar

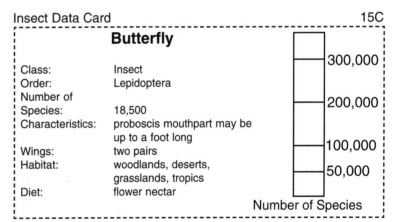

300,000
200,000
100,000
50,000
Number of Species

Mapping Monarch Migration 15D   Mapping Monarch Migration 15E

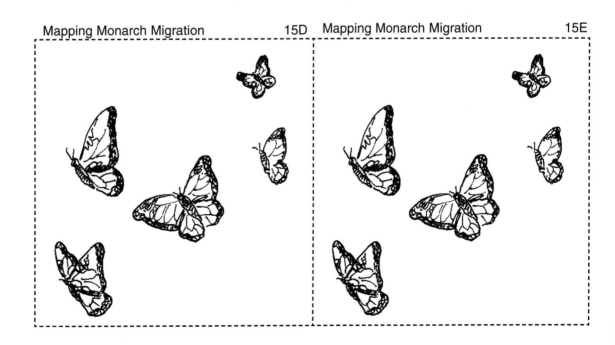

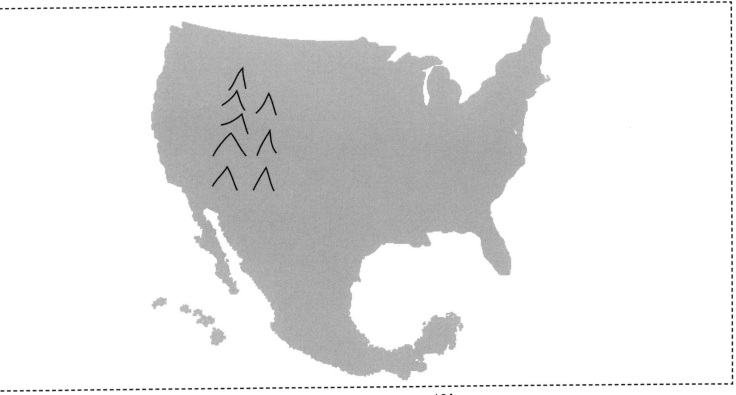

Types of Ants

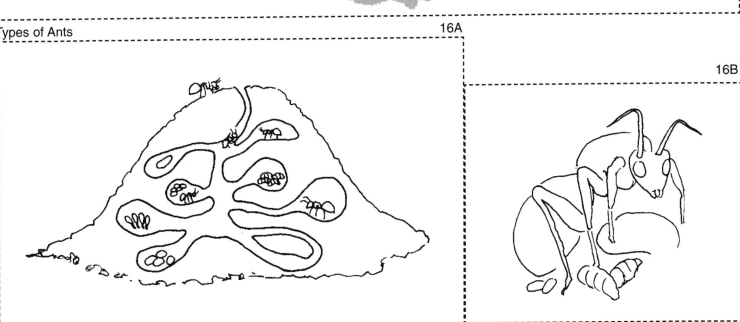

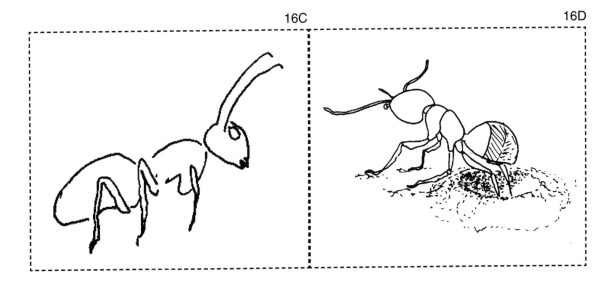

"The Termite" 16E

The Termite
by
Ogden Nash

Some primal termite knocked on wood
And tasted it, and found it good,
And that is why your Cousin May
Fell through the parlor floor today.

Types of Honeybees 17A

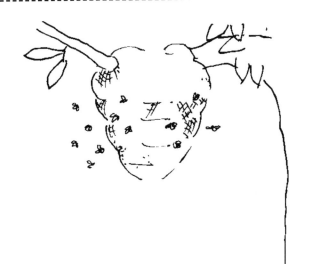

17B

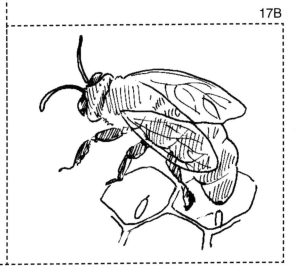

17C 17D

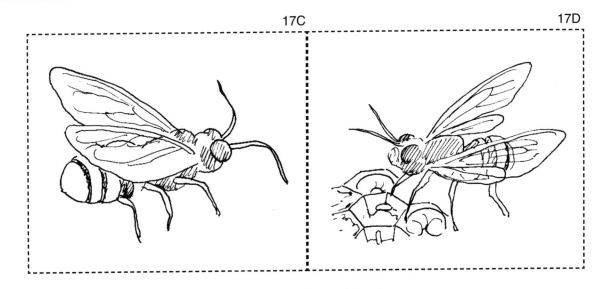

Paper Insect　　　　　　　　17E

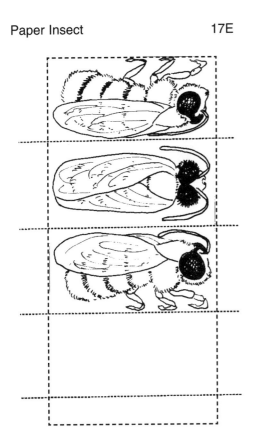

Insect Data Card　　　　　　　17F

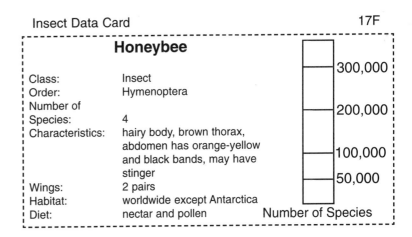

Honeybee

Class:	Insect
Order:	Hymenoptera
Number of Species:	4
Characteristics:	hairy body, brown thorax, abdomen has orange-yellow and black bands, may have stinger
Wings:	2 pairs
Habitat:	worldwide except Antarctica
Diet:	nectar and pollen

Number of Species

18A　　　　　　　　　　　　　　　18B

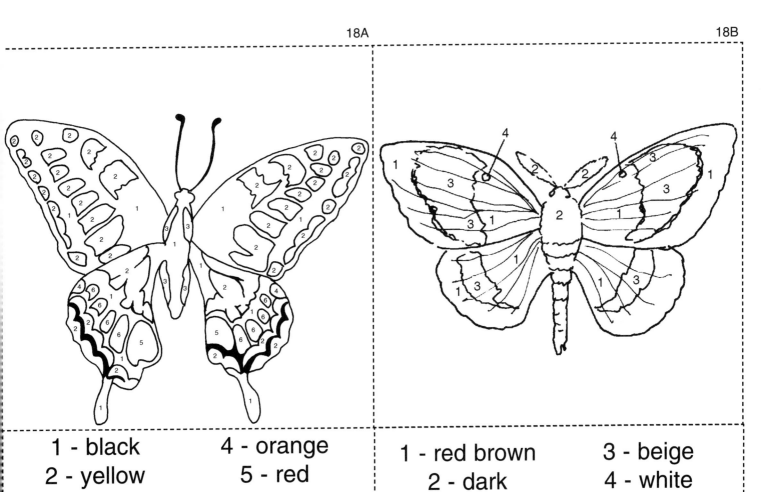

1 - black　　　4 - orange
2 - yellow　　 5 - red
3 - yellow ocre　6 - royal blue

1 - red brown　　3 - beige
2 - dark　　　　 4 - white

Paper Insect 19A

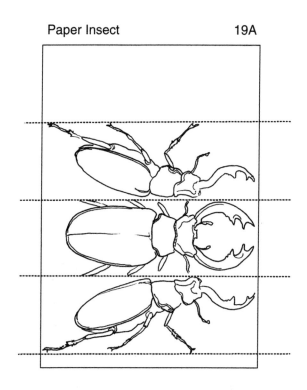

Insect Data Card 19B

Beetle

Class: Insect
Order: Coleoptera
Number of
Species: 350,000
Characteristics: usually 1/4" to 6" inches long
Wings: two pair: elytra (hard front wings) cover thin back wings; some are wingless
Habitat: deserts, forests, freshwater, saltwater
Diet: roots, leaves, flowers, fruit, dung, dead animals

300,000
200,000
100,000
50,000

Number of Species

Beetles 19C

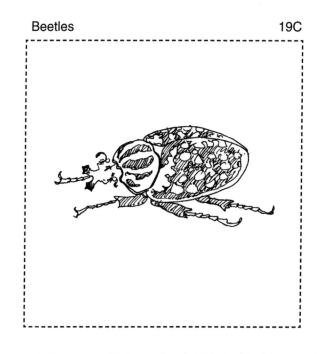

Beetles 19D

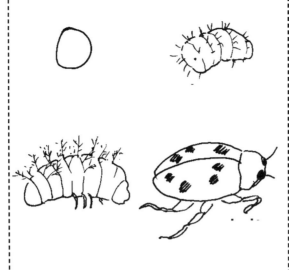

19E

19F

Helpful/Harmful Insects 20A

20B

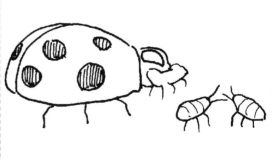

20C Aphids　　Lab 20-1

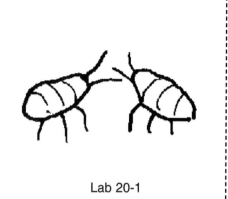

Lab 20-1

Lobsters and Crustaceans　　21A

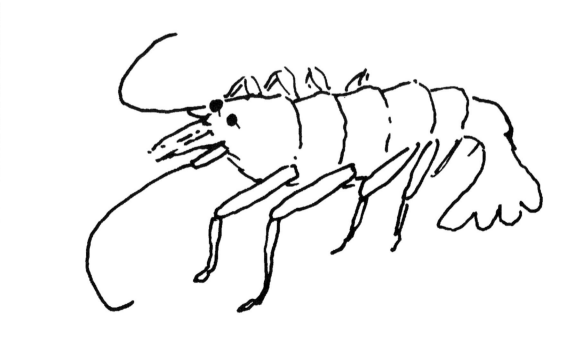

Arachnids and Insects 22A

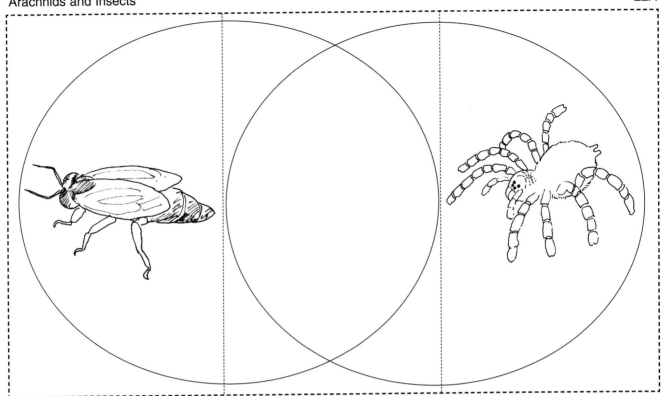

Paper Arachnid 22B

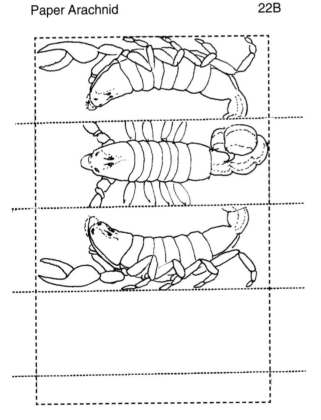

Arachnid Data Card 22C

Scorpion

Class: Arachnid
Order: Scorpionida
Number of
Species: 1,200
Characteristics: 1-8 inches long, elongated body, segmented tail with stinger
Habitat: warm, tropical regions, grasslands, savannah, forests

300,000
200,000
100,000
50,000

Number of Species

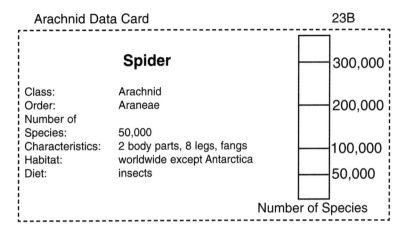

Arachnid Data Card 23B

Spider

Class: Arachnid
Order: Araneae
Number of
Species: 50,000
Characteristics: 2 body parts, 8 legs, fangs
Habitat: worldwide except Antarctica
Diet: insects

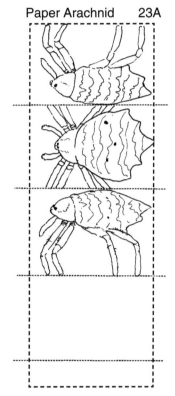

Paper Arachnid 23A

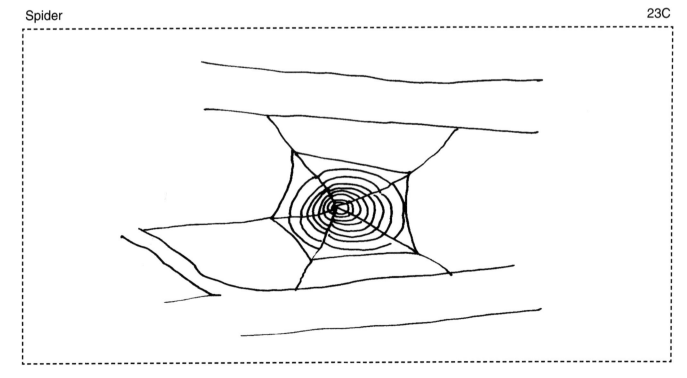

Spider 23C

Unsticky Spider Feet   Lab 23-1

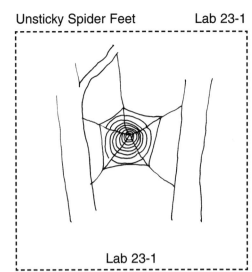

Lab 23-1

Ticks   24A

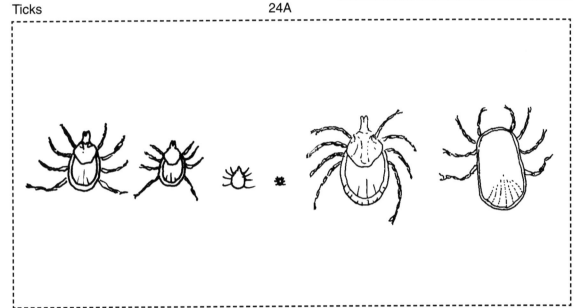

Paper Arachnid   24B

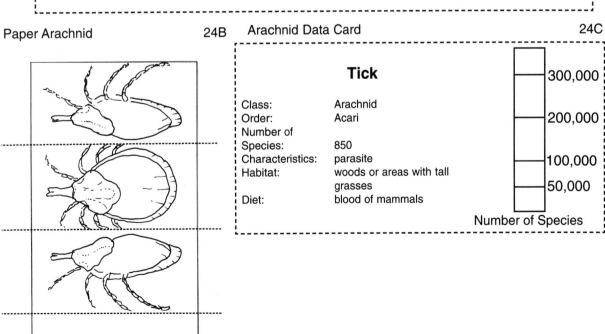

Arachnid Data Card   24C

Tick

Class:	Arachnid
Order:	Acari
Number of Species:	850
Characteristics:	parasite
Habitat:	woods or areas with tall grasses
Diet:	blood of mammals

300,000
200,000
100,000
50,000

Number of Species